4\1

Computational Colour Science using MATLAB

Computational Colour Science using MATLAB

Stephen Westland

School of Design,
University of Leeds, UK

Caterina Ripamonti

Department of Psychology,
University of Pennsylvania, USA

John Wiley & Sons, Ltd

This publication is designed to provide accurate and authoritative information with regard to the subject
matter covered. It is sold on the understanding that the Publisher is not engaged in rendering professional
services. If professional advice or other expert assistance is required, the services of a competent
professional should be sought.

Other Wiley Editorial Offices

John Wiley & Sons Inc., 111 River Street, Hoboken, NJ 07030, USA

Jossey-Bass, 989 Market Street, San Francisco, CA 94103-1741, USA

Wiley-VCH Verlag GmbH, Boschstr. 12, D-69469 Weinheim, Germany

John Wiley & Sons Australia Ltd, 33 Park Road, Milton, Queensland 4064, Australia

John Wiley & Sons (Asia) Pte Ltd, 2 Clementi Loop #02-01, Jin Xing Distripark, Singapore 129809

John Wiley & Sons Canada Ltd, 22 Worcester Road, Etobicoke, Ontario, Canada M9W 1L1

Wiley also publishes its books in a variety of electronic formats. Some content that appears in print may
not be available in electronic books.

Library of Congress Cataloging-in-Publication Data

Westland, Stephen.
  Computational colour science using MATLAB / Stephen Westland, Caterina Ripamonti.
  ———p. cm.
  Includes bibliographical references and index.

  ISBN 0-470-84562-7 (Cloth : alk. paper)
  1. Colorimetry. 2. MATLAB. I. Ripamonti, Caterina. II. Title.
  QC495.8 .W47 2004
  535.6′0285–dc22
                                                                            2003021483

British Library Cataloguing in Publication Data

A catalogue record for this book is available from the British Library

ISBN 0 470 84562 7 hardback

Typeset in 10½/13pt Times by Dobbie Typesetting Ltd, Tavistock, Devon
Printed and bound in Great Britain by TJ International Ltd, Padstow, Cornwall
This book is printed on acid-free paper responsibly manufactured from sustainable forestry
in which at least two trees are planted for each one used for paper production.

To our parents

Contents

Acknowledgements   xi

1   **Introduction**   1
    1.1   Who this book is for   1
    1.2   Why base this book upon MATLAB?   2
    1.3   A brief review of the CIE system of colorimetry   4

2   **Linear Algebra for Beginners**   13
    2.1   Some basic definitions   13
    2.2   Solving systems of simultaneous equations   14
    2.3   Transposes and inverses   16
    2.4   Linear and non-linear transforms   16

3   **A Short Introduction to MATLAB**   19
    3.1   Matrix operations   20
    3.2   Computing the transpose and inverse of matrices   22
    3.3   M-files   25
    3.4   Using functions in MATLAB   25

4   **Computing CIE Tristimulus Values**   27
    4.1   Introduction   27
    4.2   Standard colour-matching functions   27
    4.3   Interpolation methods   29
    4.4   Extrapolation methods   33
    4.5   Tables of weights   34
    4.6   Correction for spectral bandpass   35
    4.7   Chromaticity diagrams   35
    4.8   Implementations and examples   37
          4.8.1   Spectral bandpass correction   37
          4.8.2   Reflectance interpolation   39
          4.8.3   Computing tristimulus values   41
          4.8.4   Plotting the spectral locus   45

5   **Computing Colour Difference**   49
    5.1   Introduction   49
    5.2   CIELAB and CIELUV colour space   50

	5.3	CIELAB colour difference	52
	5.4	Optimised colour-difference formulae	55
		5.4.1 CMC(l:c)	55
		5.4.2 CIE94	56
		5.4.3 CIEDE2000	57
	5.5	Implementations and examples	58
		5.5.1 Computing CIELAB and CIELUV coordinates	58
		5.5.2 Computing colour difference	68

6   Chromatic-adaptation Transforms and Colour Appearance   81

	6.1	Introduction	81
	6.2	CATs	82
		6.2.1 CIECAT94	86
		6.2.2 CMCCAT97	89
		6.2.3 CMCCAT2000	90
	6.3	CAMs	92
		6.3.1 CIECAM97s	93
		6.3.2 CMCCAM2000	96
	6.4	Implementations and examples	96
		6.4.1 CATs	96
		6.4.2 Computing colour appearance	104

7   Characterization of Computer Displays   111

	7.1	Introduction	111
	7.2	Gamma	112
	7.3	The GOG model	112
	7.4	Device-independent transformation	114
	7.5	Typical characterization procedure	115
	7.6	Implementations and examples	116

8   Characterization of Cameras   127

	8.1	Introduction	127
	8.2	Correction for non-linearity	128
	8.3	Device-independent representation	129
	8.4	Implementations and examples	130

9   Characterization of Printers   141

	9.1	Introduction	141
	9.2	Physical models	142
	9.3	Neural networks	143
	9.4	Characterization of half-tone printers	145
		9.4.1 Correction for non-linearity	145
		9.4.2 Device-independent representation	146
		9.4.3 Kubelka–Munk model	147

9.5   Implementations and examples   150
       9.5.1   Half-tone printer   150
       9.5.2   Continuous-tone printer   155

10 Multispectral Imaging   **163**
10.1   Introduction   163
10.2   Computational colour constancy and linear models   164
10.3   Surface and illuminant estimation algorithms   170
10.4   Techniques for multispectral imaging   171
       10.4.1   The Hardeberg method   171
       10.4.2   The Imai and Berns method   172
       10.4.3   Methods based on maximum smoothness   172
10.5   Implementations and examples   172
       10.5.1   Deriving a set of basis functions   172
       10.5.2   Representation of reflectance spectra in a linear model   176
       10.5.3   Estimation of reflectance spectra from tristimulus values   179
       10.5.4   Estimation of reflectance spectra from camera responses   183
       10.5.5   Fourier operations on reflectance spectra   185

11 Colour Toolbox   **189**
11.1   *cband.m* (Box 1)   189
11.2   *pinterp.m* (Box 2)   189
11.3   *r2xyz.m* (Box 3)   190
11.4   *plocus.m* (Box 4)   190
11.5   *xyz2lab.m* (Box 5)   190
11.6   *lab2xyz.m* (Box 6)   190
11.7   *xyz2luv.m* (Box 7)   191
11.8   *car2pol.m* (Box 8)   191
11.9   *pol2car* (Box 9)   191
11.10 *cielabde.m* (Box 10)   191
11.11 *dhpolarity* (Box 11)   192
11.12 *cmcde.m* (Box 12)   192
11.13 *cie94de.m* (Box 13)   192
11.14 *cie00de.m* (Box 14)   193
11.15 *cmccat97.m* (Box 15)   193
11.16 *cmccat00.m* (Box 16)   193
11.17 *ciecam97s.m* (Box 17)   194
11.18 *gogtest.m* (Box 18)   194
11.19 *compgog.m* (Box 19)   195
11.20 *rgb2xyz.m* (Box 20)   195
11.21 *xyz2rgb.m* (Box 21)   195
11.22 *compigog* (Box 22)   195
11.23 *getlincam.m* (Box 23)   196

11.24 *lincam* (Box 24)   196
11.25 *gettrc* (Box 25)   196
11.26 *r2xyz* (Box 26)   197

References   **199**
Index   **205**

Acknowledgements

This book makes extensive use of the MATLAB program, which is distributed by The Mathworks, Inc. We are grateful to The Mathworks for permission to include extracts of this code.

For MATLAB product information, please contact:

The MathWorks, Inc.
3 Apple Hill Drive
Natick, MA, 01760-2098 USA
Tel: 508-647-7000
Fax: 508-647-7101
E-mail: info@mathworks.com
Web: www.mathworks.com

A user with a current MATLAB license can download trial products from the above Web site. Someone without a MATLAB license can fill out a request form on the site, and a sales rep will arrange the trial for them.

1

Introduction

1.1   Who this book is for

The growing importance of colour science in manufacturing industry has resulted in the availability of many excellent textbooks: existing texts or review papers describe the history and development of the Commission Internationale de l'Eclairage (CIE) system (Wyszecki and Stiles, 1982; Hunt, 1998), the prediction of colour difference (McDonald, 1997a; Berns, 2000; Luo, 2002a) and colour appearance (Fairchild, 1998), the relationship of the CIE system to the human visual system (Wandell, 1995; Kaiser and Boynton, 1996), and applications of colour science in technology (Green and MacDonald, 2002). However, the field of colour science is becoming ever more technical and although practitioners need to understand the theory and practice of colour science they also need guidance on how to actually compute the various metrics, indices and coordinates that are useful to the practising colour scientist. The purpose of this book is to describe methods and algorithms for actually computing colorimetric parameters and for carrying out applications such as device characterization, transformations between colour spaces and computation of various indices such as colour differences. A reasonable understanding of the main principles of the CIE system is therefore assumed, although a revision aid is provided in Section 1.3 in the form of a brief review of the CIE system of colorimetry. The reader who wishes to explore the theoretical and historical backgrounds of the topics covered by this book is encouraged to review the alternative texts mentioned above and referred to within this text. We anticipate that computer programmers, colour-image engineers and students of colour science will find this book and the associated MATLAB code useful, but hope that anyone with an interest in colour science will find the book enjoyable and informative.

Computational Colour Science Using MATLAB. By Stephen Westland and Caterina Ripamonti.
© 2004 John Wiley & Sons, Ltd: ISBN 0 470 84562 7

1.2   Why base this book upon MATLAB?

This book describes algorithms and mathematical procedures in colour science and illustrates these procedures using the numerical software tool called MATLAB. MATLAB provides several features that make it suitable for the implementation of algorithms in general, and colour-science algorithms in particular, and results in code that is easily understandable by the reader with relatively little experience of writing software. These features include the use of operations upon vectors and matrices to enable compact code that avoids the excessive use of looping procedures, the provision of a massive library of functions that the MATLAB programmer can call upon and the ease of use of graphics functions to enable the user to easily and effectively visualise complex data structures.

Most computer languages are very dependent upon a variety of 'looping' procedures to execute summations or to implement iterative techniques whereas MATLAB enables these types of operations to be performed with a fraction of the code. For example, if we have two variables x and y, each consisting of five entries, and we wish to compute the product of the corresponding entries and then sum the results to yield a single number, we might write code that in BASIC looks like the following:

```
sum = 0
FOR i =1 TO 5
sum = sum + x ( i ) *y ( i )
NEXT i
```

In MATLAB the four lines of BASIC code shown could be replaced by the single line

```
sum = x*y ;
```

Expressed in terms of linear algebra MATLAB will perform the inner product of the vectors x and y automatically. In the MATLAB environment it is not necessary to specify how many entries the variables contain, so long as the dimensions of these variables define a valid matrix operation. A variable in MATLAB can represent a single number, a vector or a whole matrix. The operation given, for example, by

```
y = 2*x
```

will assign to y twice the value of x if x defines a single number, but twice the value of every element in x if x is a vector or a matrix. The compact nature of MATLAB code allows complex and sophisticated algorithms to be explained and demonstrated with clarity and accuracy. Moreover, the computation of many colorimetric terms is ideally suited to a language that expresses variables in terms of matrices and vectors. Some procedures are best explained or implemented using loops, however, and for these situations MATLAB provides *for* and *while* looping structures.

The second strength of MATLAB is that it includes an encyclopaedic collection of subprograms, called M-files, for the solution of nearly any numerical problem. Although this book is not principally concerned with generic numerical analysis, but rather with particular colorimetric algorithms, the M-files that are available with MATLAB are useful for many computations in colour science. MATLAB provides many functions (such as those with the ability to invert matrices) and if it was necessary to spend time explaining these in detail or writing code to implement them it would detract from the main focus of this book, which is colour science. The reader may wish to refer to other textbooks (e.g. Press *et al.*, 1993) that address implementations of numerical analysis in programming languages such as C if they wish to convert the code in this book into other programming languages.

MATLAB's most spectacular feature is its capability to display graphics. Two- or three-dimensional graphs are easily constructed by even a novice MATLAB user. Thus

```
x = [1 2 3 4 5];
y = [3 5 7 9 11];
plot(x,y)
```

is sufficient code to plot a graph of the five values in the vector y against those in the vector x. Experienced programmers will find it trivial to construct sophisticated and informative graphs, and the ability to almost effortlessly visualize data is one of main advantages of using MATLAB in a research environment. MATLAB allows the user to answer complex 'what if?' questions with just a handful of code lines.

MATLAB can be confusing, however, for new users who do not have a reasonable understanding of linear algebra. For this reason, Chapter 3 provides a gentle introduction to MATLAB and Chapter 2 provides a basic introduction to linear algebra and the notation that is used throughout this book. Where possible the code that is presented has been written for clarity rather than for efficiency or speed of computation to allow the reader to understand the computational principles involved and to be able to implement them in a wide variety of programming languages. In general, special MATLAB commands have been avoided, even though their use may have made the code more efficient, to reduce the effort that would be required to translate the code into a language such as C or C++. One exception, however, is the backslash operator, which is described in Chapter 3.

Programmers who wish to use languages other than MATLAB may wish to create their own version of the backslash operator in order to easily translate the code within this book. All the MATLAB code contained within this book can be downloaded from http://www.colourware.co.uk/compute/ and from http://www.mathworks.com/matlabcentral/.

1.3   A brief review of the CIE system of colorimetry

Light is a term that we use to describe that range of wavelengths (approximately 380–780 nm) of electromagnetic radiation to which the human visual system is sensitive. When we observe the light reflected from surfaces in a scene, or when we look directly at the light emitted by light sources, we experience the sensation of colour. Colour is just one attribute of a complex and not fully understood set of properties that define the appearance of the world. Surfaces interact with light in a complex and varied way that includes the processes of absorption, scattering, refraction and diffraction, but it is the light that is reflected by the surfaces in a scene that we use to identify those surfaces by their colour. The reflectance properties of surfaces can be defined by the spectral reflectance factors that are normally measured at regular intervals in the visible spectrum of radiation. Typical reflectance spectrophotometers are able to measure the reflectance factors at intervals of 10 nm in the range 400–700 nm (though some instruments extend their measurement to shorter or longer wavelengths). Reflectance factors are normally in the range 0–1 and represent the proportional amount of light reflected in each wavelength interval. The light that we see when we look at a point in a scene clearly depends upon the spectral power distribution of the illuminating source and the reflectance properties of the surface at that point. Our visual systems detect the reflected light using the light-sensitive sense organs or retinas that form the inner lining of the back of the eyeball. Light enters the eye through the pupil and is focused onto the retina by the lens. The human retina consists of a mosaic of specialized cells called rods and cones that contain pigments that respond to light. The chemical changes that take place when the visual pigments in the rods and cones absorb light initiate electrical impulses that are subsequently processed by a neural network of brain cells and which eventually lead to the excitation of cells in various specialized areas of the outer region of the brain known as the cortex. It is still unknown where in the brain colour perception actually occurs, if indeed it occurs in any localized area, but activity in the visual cortex at the back of the brain is strongly implicated. The properties of the visual system have been reviewed elsewhere (e.g. Roberts, 2002) and only a minimal summary of the retinal processes is now presented before methods for the measurement of colour are outlined.

The rods are responsible for our vision at low levels of illumination, referred to as scotopic or night vision. At higher or photopic levels of illumination colour vision is mediated by the responses of the cones, of which there are three types, with sensitivities peaking at 420 nm (short wavelengths), 530 nm (medium wavelengths) and 560 nm (long wavelengths), termed S, M and L cones, respectively (Bowmaker, 2002). The three classes of cones are not distributed evenly throughout the retina (Williams *et al.*, 1981). In the central or foveal region, for example, only L and M cones are present and there are approximately twice as many L cones as there are M cones. The S cones are rare throughout the retina but are more concentrated in a ring around the fovea. The retina contains several layers of cells and the signals generated by the transduction of light into chemical and electrical energy in the cones activate the bipolar and ganglion cells before leaving the eye via the optic nerve. The cones and rods each provide a univariant response (Wandell, 1995; Westland, 2002) and the consequence of this is that the individual classes of cones and rods are colour blind. That is, the scotopic visual system can only perceive shades of grey and the three classes of cones considered separately are also incapable of wavelength discrimination. At least two classes of cones are required for colour vision. The photopic visual system achieves colour vision by analysing the relative responses of the three classes of cones in the eye.

The CIE developed a system for the specification of colour stimuli that was recommended for widespread use in 1931. The most important principle that allowed this development was additive colour mixing. Thus, all colour stimuli can be matched by the additive mixture of three appropriately chosen primaries. It had long been recognized that the amounts or intensities of the primaries required to match a given stimulus effectively form a specification of the stimulus in terms of the primaries that are used. The amounts of the primaries used for any given stimulus are commonly known as the tristimulus values. It is possible to determine the tristimulus values for any given stimulus using a device known as a split-field or bipartite colorimeter. In such a device an observer views a bipartite field. On one side of the field the stimulus is displayed; on the other side the additive mixture of the three primaries is displayed. The observer adjusts the intensities of each of the three primaries until the additive mixture is indistinguishable from the stimulus. Under the matching condition the field appears uniform to the observer and the tristimulus values can be read off from the device and recorded.

The measurement of the colour-matching functions by observers was a critical feature in the development of the 1931 CIE system of colorimetry since it allowed the computation of the tristimulus values for a known stimulus without the need to view the stimulus in a bipartite colorimeter. The colour-matching functions are the amounts of three primaries required to match one unit of intensity of a single wavelength of light, and were recorded for small wavelength intervals throughout the visible spectrum. If red, green and blue primaries are used and these are denoted by the symbols [R], [G] and [B], and the tristimulus values are represented by the symbols R, G

and *B*, then it is possible to write an equation to denote the matching condition:

$$S \equiv R[\text{R}] + G[\text{G}] + B[\text{B}]. \tag{1.1}$$

In this equation the symbol $\equiv$ means 'is matched by' and the stimulus is denoted by *S*. If the tristimulus values are measured separately for each wavelength in the visible spectrum, then we obtain the tristimulus values as functions of the wavelength λ: $R(\lambda)$, $G(\lambda)$ and $B(\lambda)$. These three functions of wavelength are called colour-matching functions. The additivity and linearity of colour matches allow an important property: if a stimulus S_1 is matched by R_1, G_1 and B_1 and a stimulus S_2 is matched by R_2, G_2 and B_2, then it is possible to predict in advance the tristimulus values that define a match to the stimulus defined by the additive mixture $S_1 + S_2$. Thus we can simply write

$$S_1 + S_2 \equiv (R_1 + R_2)[\text{R}] + (G_1 + G_2)[\text{G}] + (B_1 + B_2)[\text{B}]. \tag{1.2}$$

Since any real stimulus can be considered to be the sum of energy at many different wavelengths, it is possible to predict the tristimulus values for any stimulus in a similar way (without having to resort to physically determining a visual match for that stimulus using a bipartite colorimeter) given that the colour-matching functions are known.

In fact, experiments were carried out prior to the publication of the CIE system by two groups of workers, headed by Wright in 1929 and Guild in 1931, to determine colour-matching functions. The two groups of workers used different primaries and consequently the two sets of colour-matching functions were different. This raises an interesting question: Are the colour-matching functions arbitrary, given that there is a very wide choice in the selection of the primaries? Certainly, the actual tristimulus values obtained for a given stimulus are arbitrary in that they would be different if a different set of primaries was chosen. However, the matching condition is valid no matter which primaries are selected subject to some simple criteria (for example, the primaries must be independent; in other words, it must not be possible to match one of the primaries using an additive mixture of the other two, etc.). This means that if two stimuli are a visual match and are specified by the same tristimulus values under Guild's system, then they would also be a match under Wright's system. Furthermore, the two stimuli would be a match under a system defined by any other set of three primaries.

It is possible to convert tristimulus values from one system to another by a simple linear transform (see Chapter 2). It is also possible to compute the colour-matching functions for a set of known primaries given the colour-matching functions of another set of primaries. Thus, in 1931 the CIE transformed the two sets of colour-matching functions obtained from experiments carried out by Wright and Guild into a single set of colour-matching functions and reassuringly found good agreement between the two sets of data. The CIE system as we know it today is based upon a transformation of the original colour-matching functions averaged from Guild and Wright to a set of primaries known as X, Y and Z. The colour-matching functions are known for each wavelength and are therefore represented by $x(\lambda)$, $y(\lambda)$ and $z(\lambda)$.

The CIE also defined standard illuminants – tables of spectral power distributions – that can be used to compute the colour signal for a surface given the spectral reflectance factors of the surface. The introduction of tables of illuminants allowed the computation of tristimulus values for surface colours as well as for self-luminous colours. A practical formula for computing the CIE 1931 tristimulus values for a surface with spectral reflectance $P(\lambda)$ under an illuminant of relative spectral power $E(\lambda)$ is

$$X = k\Sigma E(\lambda)P(\lambda)x(\lambda),$$
$$Y = k\Sigma E(\lambda)P(\lambda)y(\lambda), \tag{1.3}$$
$$Z = k\Sigma E(\lambda)P(\lambda)z(\lambda),$$

where k is $100/[\Sigma\, y(\lambda)E(\lambda)]$.

At each wavelength interval the product $E(\lambda)P(\lambda)$ gives the amount of energy in the stimulus at wavelength λ and the amount of the primary required to match this is given by multiplying this product by the colour-matching function at that wavelength. In order to arrive at the amount of the primary required to match the stimulus it is only necessary to sum across all wavelengths [Equation (1.3)]. Note that the implication of the normalizing factor k is that the *absolute* spectral power distribution for the illuminant is not required so that, for surface colours at least, $Y = 100$ for a perfect white surface [for which $P(\lambda) = 1$ for all λ]. Furthermore, note that a perfect white surface will give $Y = 100$ for *any* illuminant E. This normalisation is reasonable given the processes of adaptation that take place in our everyday vision. In order to appreciate these processes, imagine a piece of white paper with reflectance of 1 at all wavelengths and a piece of black coal with reflectance 0.01 at all wavelengths. Now consider viewing these two surfaces indoors (under an equal-energy light source with 100 units of light at each wavelength) and outdoors (under an equal-energy light source with 10 000 units of light at each wavelength). When viewed indoors the paper reflects 100 units of light at each wavelength whereas the coal reflects only 1 unit of light at each wavelength, but the amount of light reflected outdoors is 10 000 and 100 for the paper and coal, respectively. Even though the paper reflects 100 times as much light outdoors as it does indoors, the colour appearance of the paper remains approximately constant under the two light sources. More surprisingly, the coal reflects as much light outdoors as the paper does indoors and yet the coal is veridically seen as black. This remarkable property of colour constancy is central to our whole visual experience. The normalizing factor k in the CIE system ensures that for a perfectly white surface the Y tristimulus value will always be 100 irrespective of the quantity and quality of the illuminant. One consequence of this normalization is that it is only necessary to know the relative energy of the illuminant at each wavelength.

The CIE (1931) colour-matching functions were derived from *RGB* colour-matching experiments that used a bipartite field that subtended 2° (in terms of visual angle) at the retina. A second set of colour-matching functions was measured in 1964 using a larger (10°) field size. The 1931 and 1964 colour-matching functions are based on the same *XYZ* primaries but exhibit some marked differences. One reason for this is that the distribution of cones (the light-sensitive cells in the eye) is not uniform across the retina. For example, it is known that there are no cones that contain short-wavelength-sensitive pigment in the central region of the retina known as the fovea. The present situation whereby there are two sets of colour-matching functions known as the 2-degree (1931) and the 10-degree (1964) standard observers has served the colour industry well over the last 70 years but is ultimately unsatisfactory. Users need to make a choice based upon which set of colour-matching functions best represents any given viewing situation. This presents problems from time to time when the size of the stimulus is not exactly 2° or 10°. The CIE is currently working towards the development of a set of colour-matching functions that vary continuously for a wide range of stimulus sizes.

The CIE *XYZ* tristimulus values specify a colour stimulus in terms of the visual system. It is often useful, however, to compute the chromaticity coordinates *x* and *y* from the tristimulus values:

$$x = X/(X+Y+Z), \qquad y = Y/(X+Y+Z). \tag{1.4}$$

The chromaticity diagram is derived by plotting *y* against *x* and this provides a useful map of colour space. However, it should be noted that stimuli of identical chromaticity but different luminance are collapsed onto the same point in the two-dimensional plane of the chromaticity diagram. One of the benefits of the chromaticity diagram is that, according to Grassman's law, additive mixtures of two primaries fall on a straight line joining the two points that represent the two primaries in the chromaticity diagram. If three primaries are used, then the gamut of the additive system is given by a triangle, with the vertices defined by the chromaticities of the three primaries. The gamut of all physically realizable colours is contained by the convex shape of the spectral locus and a straight line that can be considered to be drawn between the two ends of the locus. It can readily be seen that this is so if one considers any real colour stimulus to consist of the additive sum of energy at individual wavelengths.

The CIE system of colorimetry is a system of colour specification. However, it has two limitations which are important to understand. First, the system was designed for colour specification rather than for colour appearance. The chromaticities of a perfect reflecting diffuser will change as the illumination changes. However, it has already been mentioned that the colour appearance of such a surface would be expected to remain approximately constant under quite large changes in illumination. Secondly, the system is perceptually non-uniform. For a given Euclidean distance between two points in XYZ space the magnitude of the perceptual colour difference between the two stimuli represented by those points can vary by an order of magnitude or more. This second limitation in particular has presented industrial practitioners of colorimetry with serious problems, and even today not all of those problems have been resolved. Although it is useful to be able to state that two stimuli are a visual match (under the strict conditions under which the colour-matching functions were derived) if they have the same tristimulus values, it is also useful to be able to predict the visual difference between two stimuli whose tristimulus values are not identical. Ideally, we would like a uniform colour space in which equal distances in that space correspond to equal perceptual differences.

A major advance was made by the CIE in 1976 with the introduction of the CIELAB system of colour specification. This non-linear transform of the XYZ values provided partial solutions to both the problems of colour appearance and colour difference. The transformation from tristimulus values to $L*a*b*$ coordinates is given by

$$L* = 116(Y/Y_n)^{1/3} - 16,$$
$$a* = 500[(X/X_n)^{1/3} - (Y/Y_n)^{1/3}], \quad (1.5)$$
$$b* = 200[(Y/Y_n)^{1/3} - (Z/Z_n)^{1/3}],$$

where X_n, Y_n and Z_n are the tristimulus values of a specified white achromatic stimulus (see Chapter 5 for the complete equations).

CIELAB provides a three-dimensional colour space where the $a*$ and $b*$ axes form one plane and the lightness $L*$ axis is orthogonal to this plane. The CIELAB transform was intended to be used for surface colours (a separate transform, CIELUV, was provided for use with self-luminous colour stimuli such as those generated using additive colour-reproduction devices) and includes several interesting features.

Firstly, the inclusion of difference signals crudely models processes that are believed to take place in the human visual system. Thus, whereas the retina initially captures responses derived from the cone spectral sensitivities, these responses are combined at an early (retinal) stage of visual processing to provide a luminance signal and two opponent signals that can be described as being yellow-blue and red-green. Similarly, CIELAB represents colour stimuli as an achromatic signal ($L*$) and two chromatic channels representing yellow-blue ($b*$) and red-green ($a*$).

Secondly, the normalization by the illuminant achieves a colour space that makes better predictions of colour appearance than the tristimulus space from which it is derived. Thus, whereas the x and y chromaticities of a perfect white surface vary with the illuminant, the CIELAB coordinates remain constant at $L* = 100$ and $a* = b* = 0$. CIELAB also allows the representation of a colour stimulus by dimensions of lightness, chroma and hue and it is therefore reasonable to describe CIELAB as a colour-appearance space, whereas this label is not appropriate for tristimulus space which is strictly only for colour specification. However, if predictions of colour constancy using CIELAB are compared with empirical measurements of colour constancy, then it is found that the predictions are quite poor in general. The field of colour appearance has been actively researched over the last decade in particular and several advanced colour spaces (e.g. CIECAT94 and CIECAM97s) are now available for predicting colour appearance.

Thirdly, the non-linear transform of tristimulus values in the CIELAB equations allows the Euclidean distance between two points in the new space to better predict the visual colour difference between the colour stimuli represented by those two points. Consequently, the colour difference metric known as ΔE_{ab}^* and computed by the formula

$$\Delta E_{ab}^* = [(\Delta L^*)^2 + (\Delta a^*)^2 + (\Delta b^*)^2]^{1/2}, \tag{1.6}$$

where ΔL^*, for example, denotes the difference in L^* between the two samples, has been used effectively to quantify colour difference in a wide range of industries. The values of ΔL^*, Δa^* and Δb^* are given by

$$\Delta L^* = L_1^* - L_2^*$$

$$\Delta a^* = a_1^* - a_2^*$$

and

$$\Delta b^* = b_1^* - b_2^*$$

where the subscripts refer to the two stimuli concerned.

Unfortunately, although CIELAB is more perceptually uniform than XYZ space it is still a long way from being perceptually uniform. Industrial practitioners of colour science would like to be able to apply a single tolerance on the value of ΔE_{ab}^* that defines the perceptibility or acceptability boundaries throughout colour space, but this is not possible. The last two decades of the twentieth century saw a great deal of research into the development of effective colour-difference formulae. The CMC formula (named after the Colour Measurement Committee of the Society of Dyers and Colourists) was introduced in 1983 and has been widely used in industry (Clarke et al., 1984). However, a new recommendation for colour difference was recently introduced by the CIE and is known as CIEDE2000 (Luo et al., 2001). CIEDE2000, like its predecessor CMC, is not in itself a colour space (it computes colour difference starting from differences in CIELAB space) but rather describes a method for combining and weighting the differences that is more complex, and certainly more effective, than simply measuring the Euclidean distance.

Systems that are able to better predict colour difference and colour appearance are currently active areas of research for colour scientists in academia and industry. One of the factors that is actively driving research in these areas is the need to be able to effectively communicate colour between image-capture and image-reproduction devices. The proliferation of inexpensive colour-capture and -display systems, in addition to the increasing commercial use of colour on the Internet, requires increased understanding of (and ability to predict) colour difference and colour appearance.

A wide range of readable and informative texts exist for the reader who would like to explore the background and methods of the CIE system in more detail (e.g. McDonald, 1997a; Hunt, 1998; Berns, 2000).

2

Linear Algebra for Beginners

2.1 Some basic definitions

A matrix is a rectangular array of numbers and the numbers in the array are called the entries in the matrix. A two-dimensional matrix with one dimension equal to 1 is sometimes called a row matrix (a matrix with only one row) or a column matrix (a matrix with only one column). A matrix with both dimensions equal to 1 is simply a single number which we can also call a scalar. It is conventional to denote matrices by boldface upper-case symbols and row or column vectors by lower-case symbols. So, for example, the matrix $\mathbf{A}$, where

$$\mathbf{A} = \begin{bmatrix} 1 & 0 \\ 0 & 1 \end{bmatrix}$$

is a 2×2 matrix with four entries. Since only the diagonal entries (from top left to bottom right) are non-zero we can state that $\mathbf{A}$ is a diagonal matrix (furthermore, a diagonal matrix whose diagonal entries are all 1 is also called an identity matrix).

Two matrices are defined to be equal if they have the same size and their corresponding entries are equal. If $\mathbf{A}$ and $\mathbf{B}$ are matrices of the same size, then the sum $\mathbf{A} + \mathbf{B}$ is the matrix obtained by adding the entries of $\mathbf{B}$ to the corresponding entries of $\mathbf{A}$, and the difference $\mathbf{A} - \mathbf{B}$ is the matrix obtained by subtracting the entries of $\mathbf{B}$ from the corresponding entries of $\mathbf{A}$. Only matrices of the same size can be added or subtracted. As an example, if we defined the matrix $\mathbf{B}$ by

$$\mathbf{B} = \begin{bmatrix} 1 & 2 \\ 3 & 4 \end{bmatrix},$$

then we can write that

$$\mathbf{A} + \mathbf{B} = \begin{bmatrix} 2 & 2 \\ 3 & 5 \end{bmatrix}$$

and that

Computational Colour Science Using MATLAB. By Stephen Westland and Caterina Ripamonti.
© 2004 John Wiley & Sons, Ltd: ISBN 0 470 84562 7

$$\mathbf{A} - \mathbf{B} = \begin{bmatrix} 0 & -2 \\ -3 & -3 \end{bmatrix}.$$

If $\mathbf{A}$ is an $m \times r$ matrix and $\mathbf{B}$ is an $r \times n$ matrix, then the product $\mathbf{AB}$ is the $m \times n$ matrix whose entries are determined as follows. To find the entry in row i and column j of $\mathbf{AB}$, single out row i from matrix $\mathbf{A}$ and column j from matrix $\mathbf{B}$, multiply the corresponding entries from the row and column together and then add up the resulting products. Thus

$$\mathbf{AB} = \begin{bmatrix} 1 & 2 \\ 3 & 4 \end{bmatrix}.$$

In order to see how the entries were created in $\mathbf{AB}$, note that for the (i, j) entry where $i = j = 1$, we took the values 1 and 0 from the first row of $\mathbf{A}$ and the values 1 and 4 from the first column of $\mathbf{B}$ to yield $(1)(1) + (0)(3) = 1$. Note that multiplying matrix $\mathbf{B}$ by $\mathbf{A}$ resulted in matrix $\mathbf{AB}$, which was the same as $\mathbf{B}$. This special situation occurred because matrix $\mathbf{A}$ is the identity matrix. We can therefore note that multiplying a matrix by the identity matrix is like multiplying a scalar by unity. It should also be clear that a matrix $\mathbf{A}$ may only be multiplied by a matrix $\mathbf{B}$ if the number of columns in $\mathbf{A}$ is equal to the number of rows in $\mathbf{B}$.

2.2 Solving systems of simultaneous equations

Imagine that we wish to solve a problem where we need to find the values of two variables, x and y, and we are given knowledge of two relationships between the two variables. For example, we might be told that the sum of the two variables is 6 and the difference between the two is 3. We can represent this problem by a pair of simultaneous equations:

$$\begin{aligned} 6 &= x + y, \\ 3 &= x - y. \end{aligned} \tag{2.1}$$

Many readers will be familiar with this sort of problem and will possess the algebraic skills to rearrange these two equations into a form that enables one of the variables to be eliminated. In this trivial example, we can simply add the two equations together to give

$$9 = 2x,$$

from which it is now obvious that $x = 4.5$ and (by subsequent substitution) that $y = 1.5$.

However, it is often convenient to represent the problem in matrix form. The two simultaneous (or coupled) linear equations can be written as a single matrix equation of the form

$$\begin{bmatrix} 6 \\ 3 \end{bmatrix} = \begin{bmatrix} 1 & 1 \\ 1 & -1 \end{bmatrix} \begin{bmatrix} x \\ y \end{bmatrix}.$$

We can further simplify the notation by writing

$$\mathbf{a} = \mathbf{Mp}, \tag{2.2}$$

where **a** and **p** are 2×1 column matrices and **M** is a 2×2 matrix. Note that the 'inner' dimensions of the terms that are being multiplied together match; thus **Mp** is $(2 \times \underline{2})(\underline{2} \times 1)$. Matrices can only be multiplied together if their inner dimensions match in this way and matrix multiplication is sometimes referred to as computing the inner product. Note also that the dimensions of the result of computing the inner product are given by the outer dimensions. Thus, the result of a $(\underline{2} \times 2)(2 \times \underline{1})$ multiplication is a 2×1 matrix.

Matrix notation is concise and provides an alternative way to arrive at a solution to Equation (2.1). In order to solve the problem we need to compute the inverse of the matrix **M**. We denote the inverse of a matrix **M** as $\mathbf{M}^{-1}$ and define it by

$$\mathbf{I} = \mathbf{MM}^{-1},$$

where **I** is the identity matrix. Strictly, it is only possible to compute the inverse for matrices that are square. However, approximation methods can be used to compute the pseudoinverse of a non-square matrix and this procedure is denoted by the $+$ superscript symbol in this book, $\mathbf{M}^{+}$.

The identity matrix for **M** in our problem would be given by

$$\mathbf{I} = \begin{bmatrix} 1 & 0 \\ 0 & 1 \end{bmatrix}.$$

If we multiply a matrix by the identity matrix it is rather like multiplying a scalar by 1; its effect can be ignored. Thus, we can now multiply both sides of Equation (2.2) by the inverse of **M** to give

$$\mathbf{M}^{-1}\mathbf{a} = \mathbf{M}^{-1}\mathbf{Mp},$$

and since $\mathbf{M}^{-1}\mathbf{M}$ is the identity matrix we can write

$$\mathbf{p} = \mathbf{M}^{-1}\mathbf{a}$$

to give an equation that will provide a solution **p** to the simultaneous equations that were originally considered as Equation (2.1). All that is required is to be able to compute the inverse of matrix **M** and then compute the product of $\mathbf{M}^{-1}$ and **a**.

2.3   Transposes and inverses

If $\mathbf{A}$ is an $m \times n$ matrix, then the transpose of $\mathbf{A}$, denoted by $\mathbf{A}^{\mathrm{T}}$, is defined to be the $n \times m$ matrix that results from interchanging the rows and columns of $\mathbf{A}$; that is, the first column of $\mathbf{A}^{\mathrm{T}}$ is the first row of $\mathbf{A}$, the second column of $\mathbf{A}^{\mathrm{T}}$ is the second row of $\mathbf{A}$, and so forth.

If $\mathbf{A}$ is a square matrix and a matrix $\mathbf{A}^{-1}$ can be found such that

$$\mathbf{A}\mathbf{A}^{-1} = \mathbf{A}^{-1}\mathbf{A} = \mathbf{I},$$

where $\mathbf{I}$ is the identity matrix, then $\mathbf{A}$ is said to be invertible and $\mathbf{A}^{-1}$ is the inverse of matrix $\mathbf{A}$.

2.4   Linear and non-linear transforms

A linear transform is a type of function; a rule f that associates with each element in a set A one and only one element in a set B (Anton, 1994). If f associates the element b with the element a, then we write $b = f(a)$. For the most common functions, A and B are sets of real numbers, in which case f is a real-valued function of a real variable $\mathfrak{R}$. A function may associate a four-dimensional real value $\mathfrak{R}^4$ with a three-dimensional real value $\mathfrak{R}^3$, in which case we say that f is a transformation from $\mathfrak{R}^4$ to $\mathfrak{R}^3$, or that f maps $\mathfrak{R}^4$ into $\mathfrak{R}^3$. We denote this by writing $f: \mathfrak{R}^4 \to \mathfrak{R}^3$.

The simultaneous equations

$$w_1 = 2x_1 - 3x_2 + x_3 - 5x_4,$$
$$w_2 = 4x_1 + x_2 - 2x_3 + x_4,$$
$$w_3 = 5x_1 - x_2 + 4x_3,$$

define an example of a function $f: \mathfrak{R}^4 \to \mathfrak{R}^3$.

There are no squared or higher terms in this example and therefore we can further say that it is a linear transform $T: \mathfrak{R}^4 \to \mathfrak{R}^3$. In matrix form this example can be expressed as follows:

$$\begin{bmatrix} w_1 \\ w_2 \\ w_3 \end{bmatrix} = \begin{bmatrix} 2 & -3 & 1 & -5 \\ 4 & 1 & -2 & 1 \\ 5 & -1 & 4 & 0 \end{bmatrix} \begin{bmatrix} x_1 \\ x_2 \\ x_3 \\ x_4 \end{bmatrix},$$

or more efficiently as

$$\mathbf{w} = \mathbf{A}\mathbf{x}, \tag{2.3}$$

where $\mathbf{w}$ and $\mathbf{x}$ are 3×1 and 4×1 column matrices, respectively, and $\mathbf{A}$ is a 3×4 matrix. The matrix $\mathbf{A}$ is called the standard matrix for the linear transformation.

For a given standard matrix **A** it is trivial to compute **w** given **x**. However, it is not trivial to compute **x** given **w**. This situation corresponds, of course, to the problem of solving simultaneous equations and in the example given by Equation (2.3) there are four variables (the entries of the column matrix **x**) and three equations. When the number of equations is less than the number of variables we say that the system is under-determined (a system is said be over-determined if the number of equations is greater than the number of variables).

A common method for solving for **x** in Equation (2.3) is to multiply both sides of the equation by the inverse of the standard matrix which we denote $\mathbf{A}^{-1}$. However, the inverse of a non-square matrix is not defined and therefore numerical methods must be employed to compute the pseudoinverse matrix denoted by $\mathbf{A}^+$.

Recall (Section 2.2) that the product of a matrix and its inverse yields the identity matrix and therefore we can write

$$\mathbf{A}^{-1}\mathbf{w} = \mathbf{A}^{-1}\mathbf{A}\mathbf{x} \quad \text{or} \quad \mathbf{x} = \mathbf{A}^{-1}\mathbf{w}. \tag{2.4}$$

A solution for **x** is therefore possible if we can compute the inverse (or pseudoinverse) of the standard matrix **A**.

Linear algebra can be used to find mappings between one set of data and another and this is sometimes called function approximation. Now, the problem is to find the standard matrix **A** given examples of data from each of the sets. Imagine, for example, that we have a set of n camera RGB response values and we wish to find a linear transform between the corresponding n known XYZ tristimulus values. Formally, if we define **T** as the $3 \times n$ matrix of tristimulus values and **C** as the $3 \times n$ matrix of camera values, then we seek a transformation of type $T: \Re^3 \to \Re^3$ or explicitly in this case $T: \mathbf{C} \to \mathbf{T}$. Thus, we need to find the coefficients a_{11} to a_{33} for the following three equations:

$$X = a_{11}R + a_{12}G + a_{13}B,$$
$$Y = a_{21}R + a_{22}G + a_{23}B,$$
$$Z = a_{31}R + a_{32}G + a_{33}B.$$

In matrix algebra we need to find the 3×3 standard matrix **A** where

$$\mathbf{T} = \mathbf{AC}. \tag{2.5}$$

Note that if we considered each row of **A** separately for the first row we can write that

$$X = a_{11}R + a_{12}G + a_{13}B,$$

where X is the X tristimulus value, RGB are the camera values, and a_{11}, a_{12} and a_{13} are the coefficients that form the first row of **A**. If we now consider the n known samples we can write

$$\mathbf{x} = \mathbf{C}\mathbf{a}_1, \tag{2.6}$$

where $\mathbf{x}$ is the $n \times 1$ column matrix of X tristimulus values and $\mathbf{a}_1$ is a 3×1 column matrix that will be used to fill the first row of $\mathbf{A}$. Equation (2.6) represents an over-determined system for $n > 3$ since there are n simultaneous equations and three variables. Equation (2.6) can be solved by rearranging thus:

$$\mathbf{a}_1 = \mathbf{C}^{-1}\mathbf{x}, \tag{2.7}$$

where $\mathbf{C}^{-1}$ is replaced by $\mathbf{C}^+$ if $n > 3$. Although it would be possible to solve $\mathbf{A}$ for each row separately, a solution can be found directly from Equation (2.5) to yield

$$\mathbf{A} = \mathbf{C}^{-1}\mathbf{T}, \tag{2.8}$$

or, in the more likely case that $\mathbf{C}$ is a non-square matrix,

$$\mathbf{A} = \mathbf{C}^+\mathbf{T}. \tag{2.9}$$

Linear algebra can also be used to find non-linear mappings between one set of data and another. We may, for example, consider the following three equations:

$$X = a_{11}R + a_{12}G + a_{13}B + a_{14}R^2 + a_{15}G^2 + a_{16}B^2,$$
$$Y = a_{21}R + a_{22}G + a_{23}B + a_{24}R^2 + a_{25}G^2 + a_{26}B^2,$$
$$Z = a_{31}R + a_{32}G + a_{33}B + a_{34}R^2 + a_{35}G^2 + a_{36}B^2.$$

This system can again be expressed in linear algebra form as

$$\mathbf{T} = \mathbf{A}\mathbf{D}, \tag{2.10}$$

where $\mathbf{T}$ is the $3 \times n$ matrix of tristimulus values and $\mathbf{D}$ is the $6 \times n$ matrix of augmented camera values where each row contains six terms: R, G, B, R^2, G^2 and B^2. In order to define this transform we need to find the 3×6 standard matrix $\mathbf{A}$.

The solution is again achieved using

$$\mathbf{A} = \mathbf{D}^{-1}\mathbf{T}, \tag{2.11}$$

or, if $\mathbf{D}$ is a non-square matrix,

$$\mathbf{A} = \mathbf{D}^+\mathbf{T}. \tag{2.12}$$

Thus, it is evident that very similar methods can be used to determine both linear and non-linear transforms. In fact, it is reasonable to consider that the linear transform is simply a special case of a more general set of polynomial transforms.

3

A Short Introduction to MATLAB

The key to using MATLAB successfully lies in the user's ability to conceptualize data as square, rectangular, column and row matrices. Whereas most programming languages are based on ordinary algebra, whereby a symbol or name is used to represent a single numerical quantity, in MATLAB every name is assumed to be a matrix and the names can be manipulated via the rules of matrix arithmetic. MATLAB commands can be entered directly to the MATLAB Command Window at the > > prompt.

In order to illustrate the use of MATLAB let us consider the problem defined by Equation (2.2) and examine how this problem could be solved using MATLAB. To enter a 2×2 matrix called **M** we can write

```
>>M = [1 1 ; 1 - 1] ;
```

Note that the entries of **M** were entered within square brackets and that the rows were separated by a semi-colon. The final semi-colon at the end of the line is optional; if it is not present MATLAB will echo the values of **M** to the Command Window when the Return key is pressed. To enter a 2×1 column matrix **p** we would write

```
>>p = [6 ; 3] ;
```

In order to solve Equation (2.2) we need to compute the inverse of matrix **M** and then multiply this by the matrix **p**. A major feature of MATLAB is that it provides many built-in, high-level functions and the function *inv* returns the inverse of a square matrix. Thus typing

```
>>inv (M)
```

results in

Computational Colour Science Using MATLAB. By Stephen Westland and Caterina Ripamonti.
© 2004 John Wiley & Sons, Ltd: ISBN 0 470 84562 7

```
ans =
0.5000      0.5000
0.5000     -0.5000
```

and the inverse of **M** is computed and displayed in the Command Window. At any time it is possible to type the command *whos* and this displays a list of the current variables and their dimensions; thus

```
>>whos
Name   Size          Bytes       Class
M      2×2           32          double array
ans    2×2           32          double array
p      2×1           16          double array
Grand total is 10 elements using 80 bytes
```

Note that since we did not assign the output of the *inv* command to any variable it was automatically assigned to the variable ans.

We can now compute the solution to Equation (2.2) easily by typing

```
>>a = inv(M)*p
a =
4.5000
1.5000
```

which gives, of course, the same result as that achieved by the substitution method that was briefly described in Chapter 2.

In the following sections the basic properties of MATLAB are briefly introduced and some of its useful functions are described.

3.1   Matrix operations

Matrices may be added, subtracted and multiplied using the conventional symbols +, − and *. Matrices may also be easily augmented thus

```
>>M = [1 1; 1 -1];
>>M = [M; M]
M =
     1      1
     1     -1
     1      1
     1     -1
```

places a copy of the original matrix **M** below the current contents of **M** and hence produces a 4×2 matrix, whereas the command

>>M = [M M]

would produce a 2×4 matrix. Two matrices can be joined, side by side, provided that they have the same number of rows. They can also be joined one on top of the other, provided they have the same number of columns.

The colon operator is a special feature in MATLAB for constructing row vectors of evenly spaced values. The statement

>>x = 1 : 6
x =
    1   2   3   4   5   6

generates a row matrix **x** containing the integers from 1 to 6.

Individual elements of a matrix may be referenced by specifying their indices within parentheses. Thus,

>>M = [1 1 ; 1 -1] ;
x = M(1,1)
x =
        1
>>y = M(2,:)
y =
1 -1

In the preceding statement the colon operator selects the whole row. Similarly, $y = M(:,2)$ would select the whole of the second column. It is also possible to edit a single entry in a matrix by addressing it directly. Thus,

>>M = [1 1 ; 1 -1] ;
>>M(1,1) = 2 ;
M =
        2   1
        1   -1

Note that whole rows or columns can easily be selected and manipulated (copied, printed, operated upon). For example, the statement $M(1,:) = 2*M(1,:)$ would double every entry in the first row of the matrix **M**.

MATLAB provides many functions for entering and manipulating special matrices including *linspace*, *ones*, *eye*, *inv*, *length*, *diag* and *size*. As an example, the command

```
w = linspace(400,700,31);
```

would generate a 31-dimensional vector containing the values 400, 410, 420, . . . , 700 evenly spaced between 400 and 700. In order to discover the operation of other functions use the *help* command such as

```
>>help size
SIZE Size of matrix.
D = SIZE(X), for M-by-N matrix X, returns the two-element
row vector D = [M, N] containing the number of rows
and columns in the matrix. For N-D arrays, SIZE(X) returns
a 1-by-N vector of dimension lengths. Trailing singleton
dimensions are ignored.
[M,N] = SIZE(X) returns the number of rows and columns in
separate output variables. [M1,M2,M3,...,MN] = SIZE(X)
returns the length of the first N dimensions of X.
M = SIZE(X,DIM) returns the length of the dimension
specified by the scalar DIM. For example, SIZE(X,1) returns
the number of rows.
See also LENGTH, NDIMS.
```

When entering matrices in MATLAB names must begin with a letter, contain only letters or digits, and although they may be entered of any length, MATLAB only retains the first 19 characters. During a MATLAB session the values of all defined variables are stored in the workspace. The user may save the current list of variables and their associated values using the `save` command. The command `save myfile.mat`, for example, will save the workspace as a special **MATLAB** file and this may be recovered during a new session using the `load` command.

The command *clear* will remove all user-defined variables from the workspace. The format of displayed numbers during a session can be changed using the *format* command. Finally, it is important to note that **MATLAB** is case sensitive.

3.2   Computing the transpose and inverse of matrices

A matrix may be easily transposed in MATLAB using the ' operator. Thus, if **x** is a 3×1 column matrix, then the command

```
x = x';
```

will convert **x** into a 1×3 row matrix.

The *inv* operator has already been introduced for computing the inverse of a matrix when solving a pair of simultaneous equations. The *inv* command can only be used to invert matrices that are square. For non-square matrices MATLAB provides the *pinv* command that computes a pseudoinverse. Whereas the inverse of a matrix $\mathbf{A}$ is denoted by the symbol $\mathbf{A}^{-1}$, the pseudoinverse is denoted by the symbol $\mathbf{A}^{+}$.

However, it is usually more efficient and accurate (Borse, 1997) to solve systems of simultaneous equations using Gaussian elimination or, equivalently, by using MATLAB's backslash division. Thus Equation (2.2) may be solved as follows:

```
x = M\p;
```

The backslash operator is used extensively throughout this book for computing the pseudoinverse of non-square matrices (a common mistake is to confuse the backslash operator with the forwardslash operator which MATLAB uses to divide one matrix by another).

For many matrices the *inv* and *pinv* commands will generate identical results to the backslash operator. However, in some circumstances the matrix is ill-behaved. Consider the systems illustrated by the upper diagrams in Figure 3.1. The top-left diagram shows a system of two simultaneous equations for which there is an exact solution (given by the intersection of the two lines). The equations that represent these two lines are neither contradictory nor simple multiples of each other, whereas the top-right diagram shows two parallel lines for which there is no solution. The equations that represent these two lines are said to be inconsistent.

However, as the gradients of the two lines in the top-left diagram become more and more similar computation of the exact solution can become difficult and can change quite markedly with small changes in the lines themselves. A set of two equations with two unknowns is termed ill-conditioned if a small change in any of the coefficients will change the solution set from unique to either infinite or empty (Borse, 1997).

It is also interesting to consider the systems represented by the lower diagrams where two variables are represented by three equations (this is known as an over-determined system). The bottom-left diagram illustrates an over-determined system with an exact solution. However, there is no exact solution for the system represented by the bottom-right diagram but an approximate solution may be found. The system represented by the bottom-right diagram of Figure 3.1 is typical of many that are encountered during solutions to colorimetric problems.

Consider the following MATLAB commands which represent and solve a problem similar to that shown in the bottom-left diagram of Figure 3.1:

```
a = [1; 1; 2];
M = [1 -1; 1 1; 6 1];
x = M\a
```

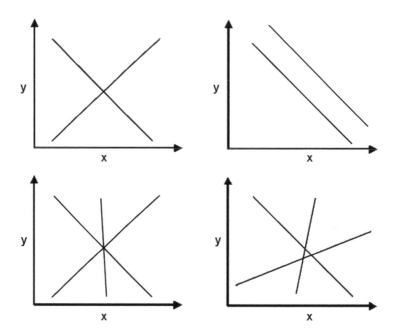

Figure 3.1 The systems of simultaneous equations illustrated graphically have an exact solution (top left and bottom left), an approximate solution (bottom right) and no solution (top right)

The solution to this problem is given by $x = (0.3846, -0.1026)$. However, if we multiply the top row of the system by a common factor, say 100, to yield the following related problem:

```
a = [100; 1; 2];
M = [100 -100; 1 1; 6 1];
x = M\a
```

then the solution reported is $x = (0.4717, -0.5282)$. Note that if we represented these two problems graphically, then they would be identical since multiplying an equation by a common factor throughout does not change it. The difference in the two solutions highlights an important property of the solution of such over-determined systems in that the solution provided by *pinv* or the backslash operator is a least-squares solution. That is, for $a = Mx$ the solution x is that which minimizes the squares of the errors between actual values of the column matrix a and predicted values of a given x. It is thus evident that multiplying one row of the system by a common factor will change the solution because it effectively changes the weight of that row in the solution. For the simple system considered the backslash and *pinv* operators generate identical solutions and this could be predicted in advance because the system is well conditioned. The condition number of a matrix is given by the MATLAB command *cond*, thus:

```
M = [1 -1 ; 1 1 ; 6 1];
cond(M)
ans =
      4.4159
```

The condition number of a matrix $\mathbf{M}$ is defined as the product of the norm of $\mathbf{M}$ and the norm of the inverse matrix $\mathbf{M}^{-1}$ (Borse, 1997). When the condition number of a matrix is high it is especially important to use the backslash operator rather than *pinv*.

3.3   M-files

A powerful property of MATLAB is that it offers the user the ability to write scripts, known as M-files. Any simple text editor such as Notepad can be used to write an M-file, but in later versions of MATLAB an Integrated Development Environment (IDE) is provided with a special MATLAB editor. Commands can be entered as a script in the same way that they would be entered into the Command Window. If the script is saved with the `.m` extension, then the commands can be executed by simply typing the name of the script. For example, an M-file called `test.m` can be executed by typing *test* in the Command Window. For some scripts it can be useful to place the command *clear* as the first line in the script so that MATLAB script is started from a clean environment. Of course, it is important to be careful to avoid using names for M-files that clash with any of MATLAB's built-in functions or M-file functions. Comments may be placed in M-files by starting the line with the % symbol.

3.4   Using functions in MATLAB

Although it is possible to create quite complicated programs using combinations of M-files (since one M-file can call another) most users will at some stage wish to create their own functions. This can also be achieved using M-files. In fact, many of the toolbox functions in MATLAB that perform some action on an arbitrary input are in fact scripts stored as M-files. In order to see how a script can be used to generate a function the following example illustrates a function called *treble* that takes a single variable as input and produces three times that variable as the output:

```
function [out] = treble(in)
out = 3*in;
```

The text for the function *treble* should be saved in an M-file called treble.m. The function is then available to the Command Window or to other M-files or functions and is simply called in any of the following ways:

```
treble(x)
y = treble(x);
[y] = treble(x);
```

The last of these formats is useful since it allows for a function to return more than one variable. Note also that the function *treble* will operate on a single number, a row or column matrix or a matrix.

A wide variety of text books (Borse, 1997; Marchand, 1999) exist for the reader who wishes to become more familiar with MATLAB before proceeding with the remainder of this book.

4

Computing CIE Tristimulus Values

4.1 Introduction

In the reproduction of colour and coloured images, trained experts known as colourists have traditionally been responsible for the assessment of the colour appearance of the colour match (Rich, 2002). Although this approach worked well for many years, in today's fast-moving global workplace more objective methods are required. Colorimetry attempts to capture the essence of colour perception and provides an objective procedure for accurate colour matching and reproduction. Tristimulus values are the basis of colorimetry and their accurate calculation is highly desired by industry for a wide range of applications. In order to compute the tristimulus values for a surface that is defined by a set of spectral reflectance values it is necessary to specify an illuminant and a set of colour-matching functions. The spectral reflectance values, the relative energy of the illuminant and the colour-matching functions must be multiplied together at each wavelength and then summed. In some cases the surface is specified at a wavelength interval that is smaller or larger than the wavelength interval of the illuminant data or the colour-matching functions. This chapter reviews methods for computing tristimulus values from spectral reflectance data and considers the use of interpolation and extrapolation where appropriate.

4.2 Standard colour-matching functions

The CIE (see Chapter 1, Section 1.3 for a brief review) originally defined the tristimulus values in terms of an integration over wavelength λ, thus:

$$X = k \int E(\lambda)P(\lambda)x(\lambda)\mathrm{d}\lambda,$$
$$Y = k \int E(\lambda)P(\lambda)y(\lambda)\mathrm{d}\lambda, \qquad (4.1)$$
$$Z = k \int E(\lambda)P(\lambda)z(\lambda)\mathrm{d}\lambda,$$

Computational Colour Science Using MATLAB. By Stephen Westland and Caterina Ripamonti.
© 2004 John Wiley & Sons, Ltd: ISBN 0 470 84562 7

where $E(\lambda)$ is the relative spectral power distribution of an illuminant, $x(\lambda)$, $y(\lambda)$ and $z(\lambda)$ are the colour-matching functions for the CIE 1931 or 1964 standard observers, $P(\lambda)$ is the spectral reflectance of a surface and k is a normalizing factor given by $100/\int E(\lambda)y(\lambda)d\lambda$. The integration was originally specified to be performed over the visible range of the electromagnetic spectrum between the wavelengths 360 nm and 830 nm. Unfortunately, analytical expressions for the colour-matching functions do not exist and so it is not possible in practice to calculate the tristimulus values according to Equation (4.1). Furthermore, the reflectance spectrum $P(\lambda)$ usually is measured at discrete intervals using commercially available reflectance spectrophotometers and is therefore also not available as an analytic expression. In 1986 the CIE adopted an alternative practice for calculating tristimulus values based upon numerical integration using wavelength intervals of 1 nm (CIE, 1986a). This leads to Equation (4.2) where the summation is carried out over the visible range of wavelengths as before:

$$X = k \sum_{360}^{830} E(\lambda)x(\lambda)P(\lambda),$$

$$Y = k \sum_{360}^{830} E(\lambda)y(\lambda)P(\lambda),\qquad\qquad(4.2)$$

$$Z = k \sum_{360}^{830} E(\lambda)z(\lambda)P(\lambda).$$

The colour-matching functions (for both the 1931 and 1964 standard observers) are provided with seven significant figures by the CIE in tabular form at 1-nm intervals between the wavelengths 360 nm and 830 nm in CIE Publication number S2 (CIE, 1986a). These are the official sets of colour-matching functions recommended by the CIE. However, for most practical applications it is suggested that an abridged set of colour-matching functions may be used at 5-nm intervals over the range 380–780 nm and these are provided in CIE Publication number 15.2 (CIE, 1986b).

The 1931 colour-matching functions are recommended whenever correlation with visual colour matching of fields of angular subtense between approximately $1°$ and $4°$ at the eye of the observer is desired. For larger angular subtenses the 1964 colour-matching functions should be used.

The use of the 5-nm colour-matching functions requires that the spectral reflectance data (for surfaces) be known at 5-nm intervals. For practical applications, the required data are often not available in an appropriate format because of abridgement (measurement at intervals greater than 5 nm) or truncation (omission of the data at the spectral extremes). Many modern reflectance spectrophotometers, for example, provide data at 10-nm intervals in the range 400–700 nm. For situations where the spectral data are abridged or truncated, the CIE recommends the use of interpolation and extrapolation, respectively (CIE, 1986b).

4.3 Interpolation methods

If reflectance data are available at 5-nm intervals, then the most accurate method to compute tristimulus values is to use the 5-nm colour-matching and illuminant data. Even if reflectance data are available at 10- or 20-nm intervals the 5-nm data can be used if interpolation methods are applied to the reflectance data. Another situation where interpolation methods may be important is where a user is computing tristimulus values for a specific non-CIE illuminant. A problem that the CIE has so far failed to solve is the disparity between illuminant spectral power distributions and light sources that serve to correspond to these illuminants. This is a particular problem with CIE illuminant D65, where although there are many lamps that are used as D65 simulators there is, in fact, no light source that replicates illuminant D65 exactly (Xu *et al.*, 2003). A practical solution to this problem is to measure the spectral power distribution of the actual light source used in a specific viewing cabinet, for example, and to use these measurements as the illuminant data in the colorimetric equations [Equation (4.2)]. This approach is sensible; unfortunately many commercial spectroradiometers provide radiance measurements at wavelength intervals of 4, 5 or 10 nm. Interpolation methods may be necessary to obtain the illuminant data at 5-nm intervals. Interpolation methods are now briefly discussed before alternative methods for computing tristimulus values are described.

A line can be drawn to fit exactly through any two points, a parabola through any three points, and an nth-degree polynomial through any $n+1$ points. Thus, if there are measurements of reflectance $P(\lambda)$ at n wavelengths an arbitrary $(n-1)$th-degree polynomial

$$P(\lambda) = a_1 \lambda^{n-1} + a_2 \lambda^{n-2} + \ldots + a_{n-1} \lambda + a_n \tag{4.3}$$

that has n coefficients can be specified by the n independent relations. A method for finding the coefficients a_1–a_n can be envisaged if we consider Equation (4.3) at each of the n wavelengths simultaneously to give the linear system

$$
\begin{aligned}
P(\lambda_1) &= a_1 \lambda_1^{n-1} + a_2 \lambda_1^{n-2} + \ldots + a_{n-1} \lambda_1 + a_n, \\
P(\lambda_2) &= a_1 \lambda_2^{n-1} + a_2 \lambda_2^{n-2} + \ldots + a_{n-1} \lambda_2 + a_n, \\
P(\lambda_3) &= a_1 \lambda_3^{n-1} + a_2 \lambda_3^{n-2} + \ldots + a_{n-1} \lambda_3 + a_n, \\
P(\lambda_4) &= a_1 \lambda_4^{n-1} + a_2 \lambda_4^{n-2} + \ldots + a_{n-1} \lambda_4 + a_n, \\
&\ldots \\
P(\lambda_n) &= a_1 \lambda_n^{n-1} + a_2 \lambda_n^{n-2} + \ldots + a_{n-1} \lambda_n + a_n,
\end{aligned}
\tag{4.4}
$$

which represents n simultaneous equations and n unknowns. In terms of linear algebra (see Chapter 2) Equations (4.4) can be efficiently represented by Equation (4.5):

$$\mathbf{p} = \mathbf{Ma}, \tag{4.5}$$

where $\mathbf{p}$ is an $n \times 1$ column matrix of reflectance values, $\mathbf{a}$ is an $n \times 1$ column matrix containing the coefficients a_1–a_n and $\mathbf{M}$ is a special $n \times n$ matrix known as the Vandermonde matrix. For a third-order polynomial, for example, the Vandermonde matrix would be constructed with the entries thus:

$$\begin{bmatrix} \lambda_1^3 & \lambda_1^2 & \lambda_1 & 1 \\ \lambda_2^3 & \lambda_2^2 & \lambda_2 & 1 \\ \lambda_3^3 & \lambda_3^2 & \lambda_3 & 1 \\ \lambda_4^3 & \lambda_4^2 & \lambda_4 & 1 \end{bmatrix}.$$

The polynomial in Equation (4.3) is referred to as the Lagrange polynomial. It is trivial to solve Equation (4.5) for the coefficients $\mathbf{a}$ using MATLAB's backslash operator: $\mathbf{a} = \mathbf{M} \backslash \mathbf{p}$ (see Chapter 3 for further information about the backslash operator). Alternatively, MATLAB also provides the functions *polyfit* and *polyval* that automatically fit and use polynomials, respectively. Thus the following code fits a fifteenth-order Lagrange polynomial to 16 reflectance values representing measurements at 20-nm intervals in the range 400–700 nm:

```
% r16 is a 1 × 16 vector containing the spectral data
w16 = linspace(400,700,16);
[P,S] = polyfit(w16,r16,15);
x = linspace(400,700,301);
y = polyval(P,x);
```

The effect of the preceding code is illustrated in Figure 4.1(a) for a typical reflectance curve. The circle symbols show the original reflectance data at 10-nm intervals. These data were directly sub-sampled (at 400, 420, 440, ..., 700 nm) to give data at 20-nm intervals and the 20-nm data were then interpolated to yield the fitted line. The solid line shows the polynomial fit illustrated at intervals of 1 nm. During the execution of the *polyfit* command MATLAB showed the warning,

```
about to call polyfit
Warning: Polynomial is badly conditioned. Remove repeated
data points or try centering and scaling as described in
HELP POLYFIT.
```

which indicates a problem with the solution of the matrix equation.

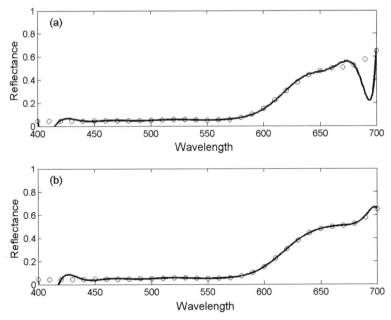

Figure 4.1 Lagrange polynomial fits (solid lines) to 20-nm reflectance data for a raw (a) and normalized (b) wavelength scale. The original data (○) were at 10-nm intervals

The problem occurs because of the construction of the Vandermonde matrix which will contain values of wavelength to the power 15 when used to fit a series of 16 reflectance values. There is clearly the likelihood of exceeding the storage capacity that MATLAB allows for a variable. The problem can be solved by modifying the code as follows so that the wavelength scale is centred and scaled. More information on this process can be found by typing *help polyfit*. The effect of the normalizing procedure can be seen by the solid line in Figure 4.1(b).

```
% r16 is a 1 × 16 vector containing the spectral data
w16 = linspace(400,700,16);
[P,S,mu] = polyfit(w16,r16,15);
x = linspace(400,700,301);
y = polyval(P,x,[],mu);
```

Note, however, that even using the normalized wavelength scale although the Lagrangian polynomial fits the 20-nm data exactly it would make quite poor estimates (particularly towards the two ends of the spectrum) if it was used to interpolate the 20-nm data to yield 10-nm intervals. This problem can be solved by using a family of polynomials (each of which fits a relatively small number of points) rather than trying to fit the whole spectrum with a single polynomial. Indeed, CIE Publication Number 15.2 recommends that interpolation be carried

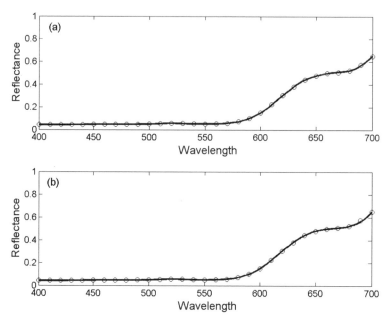

Figure 4.2 Polynomial fits (solid lines) to 20-nm reflectance data using the *interp1* function and cubic (a) and spline (b) options. The original data (○) were at 10-nm intervals

out using a third-degree polynomial from neighbouring data within twice the measurement interval (CIE, 1986b). This means that a Langrange interpolation formula should be used for four data points (two either side of the point to be interpolated). Interpolation is thus performed piecewise.

MATLAB provides some excellent interpolation functions and the most widely used is *interp1* which is used in the following way:

```
p = interp1(x,y,x1,<option>);
```

The vectors **x** and **y** are the data through which the interpolated curve must pass. The points specified in the vector **x1** are the points at which the vector **p** must be estimated. The default option is a linear interpolation in which the **y** points are simply connected by straight lines. This works surprisingly well for many reflectance curves given that the sampling interval is 20 nm or less. Other options include 'cubic' and 'spline'. The cubic fit performs cubic interpolation piecemeal in the way stipulated by the CIE. The results of interpolation to 20-nm data are shown for the cubic and spline options in Figure 4.2.

In order to evaluate the performance of interpolation techniques a set of reflectance spectra measured for 404 natural objects (Westland *et al.*, 2000) at 10-nm intervals were sub-sampled to generate data at 20-nm intervals. Interpolation techniques were then used to fit the 20-nm data and to predict the reflectance at

Table 4.1 Performance of different interpolation methods on a large set of reflectance spectra

Interpolation method	Mean rms (ΔE_{ab}^*)	Maximum rms (ΔE_{ab}^*)
Fifteenth order Lagrange	0.2865 (1.7454)	3.4720 (18.3091)
Linear	0.0198 (0.4779)	0.0600 (1.3991)
Cubic spline	0.0082 (0.0330)	0.0297 (0.1305)
Cubic Lagrange	0.0002 (0.0051)	0.0092 (0.1508)
pinterp	0.0050 (0.0128)	0.0244 (0.1487)

10-nm intervals. The original and interpolated spectra were compared and root-mean-square (rms) errors computed for each sample. Table 4.1 shows the mean and maximum rms values using several different interpolation techniques.

A general problem with Lagrange polynomials is that they are susceptible to wild oscillations and, as a consequence, can lead to large associated errors when used to interpolate data. An improvement is to successively increase the degree of the fitted polynomial, one degree at a time, and to use the change in the computed values from one step to the next as an indication of the errors. This procedure is known as Neville's algorithm. However, the results shown in Table 4.1 for a large number of reflectance spectra illustrate that the cubic Lagrange polynomial is almost certainly adequate for any practical interpolation of reflectance spectra. This may not be the case for the interpolation of spectral power distributions of light sources, however, since these can be spiky.

For users who may wish to write code in languages other than MATLAB we provide a function called *pinterp* that performs interpolation by cubic Langrange polynomials and that contains no MATLAB library calls other than the backslash operator. Table 4.1 shows that this function does not perform as well as *interp1* and so the simplicity of the code is at the cost of some accuracy. However, *pinterp* does outperform MATLAB's cubic spline interpolation and it is suggested that for most practical purposes the code would provide adequate interpolation.

Table 4.1 also shows (in parentheses) interpolation errors as CIELAB colour differences ΔE_{ab}^* computed for illuminant D65 for the 1964 observer. Using this colorimetric measure, the maximum errors for the cubic spline, cubic Lagrange and *pinterp* fits are all comparable, although on average the cubic Lagrange still performs best.

4.4 Extrapolation methods

A further problem that can occur when computing tristimulus values is that many reflectance spectrophotometers provide reflectance data in the range 400–700 nm and yet the 5-nm colour-matching functions are defined over 380–780 nm. It is possible to extrapolate the reflectance data and the method recommended by the CIE is to extend the reflectance data by using the most extreme value as an estimate of all values beyond that extreme (CIE, 1986b). So, for example, if the calculation is being carried out at 10-nm intervals and the reflectance data are in the range 400–700 nm, then values of reflectance at 710, 720, ..., 780 nm are set equal to the value at 700 nm. A similar procedure applies to the shorter wavelength. Although it could be suggested that more accurate extrapolation methods could be employed it should be remembered that extrapolation is far more dangerous than interpolation. Also, the fact that the colour-matching functions have very small values below 400 nm and above 700 nm means that the errors that result from the CIE method generally are very small and the risk of using sophisticated extrapolation techniques is not justified.

4.5   Tables of weights

Some practitioners prefer to use weighting tables where the terms $E(\lambda)x(\lambda)$, $E(\lambda)y(\lambda)$ and $E(\lambda)z(\lambda)$, as used in Equation (4.2), are pre-computed at each wavelength interval. These weighting tables can be computed from the CIE colour-matching functions and illuminants. The benefit to the user in using these tables is that Equation (4.2) can be replaced by Equation (4.6),

$$
\begin{aligned}
X &= \sum W_x(\lambda)P(\lambda), \\
Y &= \sum W_y(\lambda)P(\lambda), \\
Z &= \sum W_z(\lambda)P(\lambda),
\end{aligned}
\tag{4.6}
$$

where the weight vectors W_x, W_y and W_z also include the normalizing constant k from Equation (4.2). The CIE recommends that such tables of weighting factors should be provided for the full range of wavelengths, 360–830 nm, so that they may be used for any degree of truncation by adding the weights at the unmeasured wavelengths to those at the extreme measured wavelengths.

A set of useful weights is provided by the American Society for Testing and Materials in E308-01 (ASTM, 2001). The E308-01 tables are provided only for the range of wavelengths 360–780 nm but are suitable for most practical applications. They are provided at 10- and 20-nm intervals. The fact that the E308-01 tables are abridged to intervals of 10 and 20 nm has resulted in them probably being the most widely used method for computing tristimulus values since the 10-nm data, in particular, are suitable for direct use with reflectance data obtained from most reflectance spectrophotometers without interpolation. The ASTM publication provides the data in two main tables: ASTM Table 5 should be used with reflectance data that have been corrected for the spectral bandpass of the instrument whereas ASTM Table 6 has the spectral-bandpass correction built in and should be used with reflectance data that have not been corrected. The majority of reflectance spectrophotometers that are commercially available do not correct for the spectral bandpass of the instrument.

The ASTM tables of weights are available in hard copy or electronic form from the ASTM web site http://www.astm.org.

4.6 Correction for spectral bandpass

Figure 4.3 shows a triangular bandpass function for a typical reflectance spectrophotometer. The triangular function is effectively the spectral sensitivity of the spectrophotometer at wavelength λ_i and it can be seen that the spectrophotometer integrates energy between λ_{i-1} and λ_{i+1}. The effect of a bandpass shape as shown in Figure 4.3 is that the measured reflectance data P' need to be corrected to obtain the true reflectance data P.

Stearns and Stearns (1988) and Venable (1989) have proposed methods for spectral bandpass correction. The Stearns and Stearns correction is given by Equation (4.7),

$$P_i = -\alpha P'_{i-1} + (1 + 2\alpha)P'_i - \alpha P'_{i+1}, \tag{4.7}$$

where α is equal to 0.083 and where, if the wavelength being corrected is the first or last one in the sequence, Equation (4.8) is used,

$$P_i = (1 + \alpha)P'_i - \alpha P'_{i\pm 1}. \tag{4.8}$$

It is important to know, therefore, whether the spectral reflectance values from a given reflectance spectrophotometer have been corrected for spectral bandpass by the operating software in order that the correct tables of weights are used. The bandpass correction is not built in to the CIE 1-nm and 5-nm data and therefore if these sets of colour-matching functions are used, then it is important that the reflectance data are corrected for bandpass dependence.

4.7 Chromaticity diagrams

Chromaticity coordinates are computed from tristimulus values according to Equations (4.9),

$$\begin{aligned}
x &= X/(X + Y + Z), \\
y &= Y/(X + Y + Z), \\
z &= Z/(X + Y + Z).
\end{aligned} \tag{4.9}$$

Of course, it is evident that $x + y + z = 1$ and therefore it is usual to quote just two of the coordinates (by convention, x and y are selected) in addition to one of the tristimulus values (Y is selected because, for the 1931 observer, it is equivalent

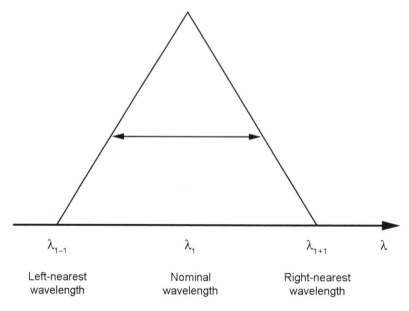

Figure 4.3   Triangular bandpass function of a typical spectrophotometer

to luminance expressed in units of cd/m^2). It is sometimes useful to be able to compute the tristimulus values from an x, y, Y specification and this can be accomplished using Equations (4.10),

$$X = xY/y,$$
$$Z = (1 - x - y)Y/y. \tag{4.10}$$

The chromaticity coordinates provide a useful representation especially for additive colour-reproduction devices where, for any luminance plane, the gamut of the device is defined by the polygon whose vertices are the chromaticities of the device primaries. Note, however, that such device gamuts are three-dimensional so, for example, for a colour monitor it will not be possible to obtain the full range of chromaticities at all luminance levels (Morovic, 2002).

The chromaticities of the spectral locus of a chromaticity diagram can be obtained directly from the tables of weights as shown in Equations (4.11),

$$x = W_x(\lambda)/[W_y(\lambda) + W_y(\lambda) + W_z(\lambda)],$$
$$y = W_y(\lambda)/[W_x(\lambda) + W_y(\lambda) + W_z(\lambda)]. \tag{4.11}$$

The weights in Equations (4.11) can be replaced by the colour-matching functions and in this case the chromaticity coordinates are computed for the appropriate observer and for the equal-energy illuminant (illuminant E). Figure 4.4 shows the spectral locus that is generated using Equation (4.11) and the tables of weights at 10-nm intervals. In order to generate a smooth spectral locus 5-nm intervals or less are required.

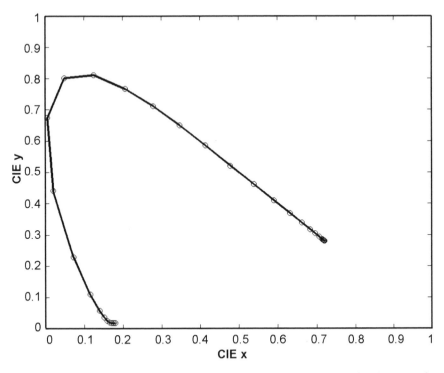

Figure 4.4 Spectral locus plot using ASTM Table 5 at 10 nm intervals for the 1964 observer and illuminant D65

4.8   Implementations and Examples

4.8.1   Spectral bandpass correction

The function *cband* applies the Stearns–Stearns correction method [Equations (4.7) and (4.8)] to a reflectance vector.

Box 1: *cband.m*

```
function [cP] = cband(P)

% function [cP] = cband(P)
% applies Stearns-Stearns spectral bandpass correction
% operates on matrix P of dimensions 1 by n
% where n is the number of wavelengths
% returns corrected matrix cP
```

```
dim = size(P);
if (dim(1) == 1) | (dim(2) == 1)
  P = P(:)'; % force to be a row matrix
else
  disp('P must be a row matrix');
  return;
end

a = 0.083;
n = length(P);
for i=2:n-1
  cP(i) = -a*P(i-1) + (1 + 2*a)*P(i) - a*P(i+1);
end
cP(1) = (1 + a)*P(1) - a*P(2);
cP(n) = (1 + a)*P(n) - a*P(n-1);
```

The format for this function is

```
[cp] = cband(p)
```

where **p** is an $n \times 1$ or $1 \times n$ matrix. In *cband* the dimensions of **p** are checked to ensure that only a single reflectance spectrum has been passed to the function. The **p** matrix is then converted to a row matrix. The MATLAB command

```
p = p(:)
```

converts the matrix **p** into a column matrix and the transpose function is added

```
p = p(:)'
```

to ensure that **p** is a row matrix.

This function operates on a single reflectance spectrum although it would be relatively easy to modify the code so that it operates on an $n \times m$ matrix of m reflectance spectra. The main purpose of this book is education rather than producing the fastest and most efficient code, and therefore most of the functions have been written to clearly demonstrate the computations involved. The following modification of *cband*, however, demonstrates how the code would be changed to allow more than one reflectance spectrum to be passed to the function:

```
function [cp] = cband2(p)

% applies Stearns-Stearns spectral bandpass correction
% operates on matrix P of dimensions n by m

a = 0.083;
dim = size(p);
n = dim(1);

for i = 2:n-1
    cp(i,:) = -a*p(i-1,:) + (1 + 2*a)*p(i,:) - a*p(i+1,:);
end
cp(1,:) = (1 + a)*p(1,:) - a*p(2,:);
cp(n,:) = (1 + a)*p(n,:) - a*p(n-1,:);
```

4.8.2 Reflectance interpolation

The CIE recommended method for interpolation of reflectance spectra is to use cubic polynomial interpolation using two points either side of the wavelength to be evaluated. If the reflectance spectrum is available at intervals of 20 nm, then the value of reflectance at 470 nm, for example, would be calculated using a cubic polynomial fitted through the reflectance at 440, 460, 480 and 500 nm. The function *pinterp* takes an *N*-dimensional reflectance vector and applies piecewise cubic polynomial interpolation to generate an additional point between each pair of points in the vector. Most reflectance spectrophotometers provide reflectance data at intervals of 10 nm and so can be used directly with the ASTM tables of weights. Some older instruments only produce data at 20-nm intervals, however, and therefore the most practical use of this function will be to reduce the sampling interval from 20 nm to 10 nm. Thus, if the input to *pinterp* is a 16-dimensional vector the output will be a 31-dimensional vector. This function therefore effectively doubles the sampling rate of the input vector.

Box 2: *pinterp.m*

```
function [s] = pinterp(p)

% function [s] = pinterp(p)
% applies interpolation to double the sampling
% rate of the n by 1 matrix p
% returns interpolated matrix s
```

```
dim = size(p);
if (dim(1) == 1) | (dim(2) == 1)
  p = p(:)'; % force to be a row matrix
else
  disp('p must be a row matrix');
  return;
end

N = length(p);

for i=1:N-1
  if (i==1)
    index1 = i;
    index2 = 1.5;
  elseif (i==N-1)
    index1 = i-2;
    index2 = 3.5;
  else
    index1 = i-1;
    index2 = 2.5;
  end

  tempy = p(index1:index1+3);
  tempx = [1 2 3 4];

  tempx = tempx(:);
  tempy = tempy(:);
  % Construct Vandermonde matrix.
  V(:,3+1) = ones(length(tempx),1);
  for j = 3:-1:1
    V(:,j) = tempx.*V(:,j+1);
  end
  % Solve least squares problem
  g = V\tempy;
  r = tempy - V*g;
  temp(i) = g(4) + g(3)*index2 + g(2)*index2*index2 +
  g(1)*index2*index2*index2;
end

for i=1:N-1
  s(i*2 - 1) = p(i);
```

```
   s(i*2) = temp(i);
 end
 s((N-1)*2+1) = p(N);
```

The format for this function is

```
[s] = pinterp(p)
```

where **p** is an $n \times 1$ or $1 \times n$ matrix. The function *pinterp* includes a special MATLAB operator called *ones* that simply creates an array of 1s that will form the rightmost column of the Vandermonde matrix [see Equation (4.5)]. Table 4.1 illustrates the performance of the *pinterp* function.

4.8.3   Computing tristimulus values

The function *r2xyz* computes *XYZ* tristimulus values using ASTM Table 5 colour-matching functions from the ASTM standard (ASTM, 2001). The reflectance data should be in the range [0, 1] rather than in per cent format and must be sampled at 10-nm intervals. Since ASTM Table 5 is used it is assumed that the reflectance data have been corrected for the spectral bandpass properties of the spectrophotometer that was used for their measurement. A typical call for a reflectance spectrum sampled at 10-nm intervals in the range 400–700 nm would be

```
[xyz] = r2xyz(p,400,700,'d65_64');
```

where **p** represents a 31×1 matrix. The second and third arguments relate to the shortest and longest wavelengths available in the reflectance data. The fourth argument specifies the illuminant and observer combination to be used.

Box 3: *r2xyz.m*

```
function [xyz] = r2xyz(p, startlam, endlam, obs)

% function [xyz] = r2xyz(p, startlam, endlam, obs)
% computes XYZ from reflectance p using a table of weights
% operates on matrix p of dimensions 1 by n for
% illuminants A, C, D50, D55, D65, D75, F2, F7, F9
```

```
% and for the 1931 and 1964 observers
% set obs to 'd65_64 for D65 and 1964, for example
% the startlam and endlam variables denote the first and
% last wavelengths (eg. 400 and 700) for your reflectance
% which must be integers of 10 in the range 360-780

if ((endlam780) | (startlam < 360) | (rem(endlam,10) ~=0) |
(rem(startlam,10) ~=0))
    disp('start and end wavelengths must be divisible by 10')
    disp('wavelength range must be 360-780 or less');
    return;
end


load weights.mat
% weights.mat contains the tables of weights
if strcmp('a_64',obs)
    cie = a_64;
elseif strcmp('a_31', obs)
    cie = a_31;
elseif strcmp('c_64', obs)
    cie = a_64;
elseif strcmp('c_31', obs)
    cie = c_31;
elseif strcmp('d50_64', obs)
    cie = d50_64;
elseif strcmp('d_50', obs)
    cie = d_50;
elseif strcmp('d55_64', obs)
    cie = d55_64;
elseif strcmp('d55_31', obs)
    cie = d55_31;
elseif strcmp('d65_64', obs)
    cie = d65_64;
elseif strcmp('d65_31', obs)
    cie = d65_31;
elseif strcmp('d75_64', obs)
    cie = d75_64;
elseif strcmp('d75_31', obs)
    cie = d75_31;
elseif strcmp('f2_64', obs)
    cie = f2_64;
elseif strcmp('f2_31', obs)
```

```
  cie = f2_31;
elseif strcmp('f7_64', obs)
  cie = f7_64;
elseif strcmp('f7_31', obs)
  cie = f7_31;
elseif strcmp('f9_64', obs)
  cie = f9_64;
elseif strcmp('f9_31', obs)
  cie = f9_31;
else
  disp('unknown option obs');
  disp('use d65_64 for D65 and 1964 observer'); return;
end

% check dimensions of P
dim = size(p);
if (dim(1) == 1) | (dim(2) == 1)
  p = p(:)'; % force to be a row matrix
else
  disp('p must be a row matrix');
  return;
end

N = ((endlam-startlam)/10 + 1);
if (length(p) ~= N)
  disp('check dimensions of p'); return;
end

% deal with possible truncation of reflectance
index1 = (startlam - 360)/10 + 1;
if (index1 > 1)
  cie(index1,:) = cie(index1,:) + sum(cie(1:index1-
  1,:));
end
index2 = index1 + N - 1;
if (index2 < 43)
  cie(index2,:) = cie(index2,:) + sum(cie(index2+
  1:43,:));
end
cie = cie(index1:index2,:);

xyz = (p*cie)*100/sum(cie(:,2));
```

```
% note that 100/sum(cie(:,2)) is the normalising factor k
% so that Y = 100 for a perfect reflecting diffuser
```

The command *load weights.mat* loads the values of ASTM Table 5 and the *whos* command would reveal the following variables:

Name	Size	Bytes	Class
a_64	43x3	1032	double array
a_31	43x3	1032	double array
c_64	43x3	1032	double array
c_31	43x3	1032	double array
d50_64	43x3	1032	double array
d50_31	43x3	1032	double array
d55_64	43x3	1032	double array
d55_31	43x3	1032	double array
d65_64	43x3	1032	double array
d65_31	43x3	1032	double array
d75_64	43x3	1032	double array
d75_31	43x3	1032	double array
f2_64	43x3	1032	double array
f2_31	43x3	1032	double array
f7_64	43x3	1032	double array
f7_31	43x3	1032	double array
f9_64	43x3	1032	double array
f9_31	43x3	1032	double array

The file *weights.mat* therefore contains weights that represent the 1964 and 1931 colour-matching functions for the CIE illuminants A, C, D50, D55, D65, D75, F2, F7 and F9. The leftmost column in the preceding list shows the valid observer/illuminant options that can be used as the fourth argument in *r2xyz*. The white points of the illuminants are given (ASTM, 2001) by Table 4.2.

Following some basic checks on the arguments to the function the issue of truncation is addressed in the *r2xyz* code. If the reflectance data are only available at 400 nm and higher, for example, then the values of the weights at wavelengths lower than 400 nm are added to the value of the weights at 400 nm. Summation of the product of the weights and the reflectance data is then performed at 400 nm and upwards. This is equivalent to extending the reflectance data below 400 nm using the value of reflectance at 400 nm. A similar process is carried out for the upper wavelength that is available. This procedure is in accordance with the CIE recommendation for dealing with truncated reflectance data.

Table 4.2 White points of illuminants used in *r2xyz.m* and other functions

	1931			1964		
	X	Y	Z	X	Y	Z
A	109.850	100.00	35.585	111.144	100.00	35.200
C	98.074	100.00	118.232	97.285	100.00	116.145
D50	96.422	100.00	82.521	96.720	100.00	81.427
D55	95.682	100.00	92.149	95.799	100.00	90.926
D65	95.047	100.00	108.883	94.811	100.00	107.304
D75	94.072	100.00	122.638	94.416	100.00	120.641
F2	99.186	100.00	67.393	103.279	100.00	69.027
F7	95.041	100.00	108.747	95.792	100.00	107.686
F9	100.962	100.00	64.350	103.863	100.00	65.607

4.8.4 Plotting the spectral locus

The following code was used to generate the rather jagged chromaticity plot shown in Figure 4.4:

```
clear

% load the ASTM tables
load weights.mat

% d65_64 is a 43 × 3 matrix
% the data at the extreme ends of the spectrum
% generate divide-by-zero and are not required
d = d65_64(4:37,:);

x = d(:,1)./(d(:,1) + d(:,2) + d(:,3));
y = d(:,2)./(d(:,1) + d(:,2) + d(:,3));

plot(x,y,'k-')

axis([0 1 0 1])
xlabel('CIE x')
ylabel('CIE y')
```

The function *plocus* returns a 59×2 matrix containing the chromaticity coordinates of the spectral locus at 5-nm intervals between 430 nm and 720 nm. A typical call would be

```
[xy] = plocus('d65_64');
```

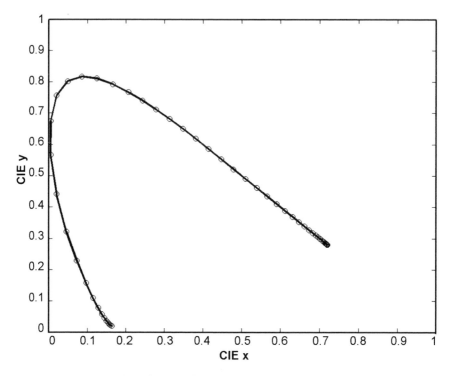

Figure 4.5 Spectral locus plot using the *plocus* function

which would provide the spectral locus for the 1964 observer using illuminant D65. The function operates by first discarding the entries for very short and very long wavelengths in the weights. This is justified because, not only do the chromaticities change very little at the extreme ends of the spectrum but the round-off errors become large relative to the small values of the weights at these wavelengths [this can result in spurious values when Equations (4.10) are used]. The abridged tables are then interpolated using the function *pinterp* to increase the sampling rate from 10 nm to 5 nm. Finally, Equations (4.11) are applied to give the chromaticity coordinates of the spectral locus (Figure 4.5). An even smoother plot of the spectral locus would be obtained if *pinterp* was called twice within the *plocus* function.

Box 4: *plocus.m*

```
function [xy] = plocus(obs)

% function [xy] = plocus(obs)
% computes spectral locus xy using interpolated ASTM
```

```
% weights
% see function r2xyz for valid values for obs
load weights.mat

if strcmp('a_64',obs)
  cie = a_64;
elseif strcmp('a_31', obs)
  cie = a_31;
elseif strcmp('c_64', obs)
  cie = a_64;
elseif strcmp('c_31', obs)
  cie = c_31;
elseif strcmp('d50_64', obs)
  cie = d50_64;
elseif strcmp('d_50', obs)
  cie = d_50;
elseif strcmp('d55_64', obs)
  cie = d55_64;
elseif strcmp('d55_31', obs)
  cie = d55_31;
elseif strcmp('d65_64', obs)
  cie = d65_64;
elseif strcmp('d65_31', obs)
  cie = d65_31;
elseif strcmp('d75_64', obs)
  cie = d75_64;
elseif strcmp('d75_31', obs)
  cie = d75_31;
elseif strcmp('f2_64', obs)
  cie = f2_64;
elseif strcmp('f2_31', obs)
  cie = f2_31;
elseif strcmp('f7_64', obs)
  cie = f7_64;
elseif strcmp('f7_31', obs)
  cie = f7_31;
elseif strcmp('f9_64', obs)
  cie = f9_64;
elseif strcmp('f9_31', obs)
  cie = f9_31;
else
  disp('unknown option obs');
```

```
   disp('use d65_64 for D65 and 1964 observer'); return;
end

% ignore the extreme wavelengths
cie = cie(8:37,:);

% interpolate to double the sampling rate
% the following three lines may be repeated
% for finer resolution
cie1(:,1) = pinterp(cie(:,1))';
cie1(:,2) = pinterp(cie(:,2))';
cie1(:,3) = pinterp(cie(:,3))';

xy(:,1) = cie1(:,1)./sum(cie1')';
xy(:,2) = cie1(:,2)./sum(cie1')';
```

5
Computing Colour Difference

5.1 Introduction

Although the system of colour specification introduced in 1931 by the CIE and augmented in 1964 has served the colour industry well, there remain a number of problems. One of the main problems is that in terms of visual perception it is very non-uniform. Equal changes in x, y or Y do not correspond to perceived differences of equal magnitude. Most attempts to develop more uniform spaces have sought to find linear or non-linear transforms of the tristimulus values or chromaticity coordinates to give a more uniform colour space. In 1976 the CIE recommended two new colour spaces for general use (CIE, 1986b): CIE $L*a*b*$ and CIE $L*u*v*$, also known as CIELAB and CIELUV. CIELUV was intended to be used to specify the colours of lights and other self-luminous sources, whereas CIELAB was intended to be used for the specification of surface colours. It is possible to compute a colour difference for two stimuli in CIELAB space by calculating the Euclidean distance in the space between the two points that represent the stimuli in the space [Equation (1.6)]. The CIELAB colour-difference formula has been used extensively for quality control in industry but its application is limited because although CIELAB space is more perceptually uniform than the tristimulus space on which it is based, it is still far from being perfectly uniform. The consequence of this is that for equal perceptual colour differences between pairs of samples, the values of CIELAB colour difference ΔE_{ab}^* computed between points representing the pairs in CIELAB space can vary by an order of magnitude. Since 1976 attempts to generate better metrics for the prediction of colour differences have concentrated on finding more sophisticated measures of distance. A summary of the developments is not given in detail here (see Smith, 1997; Berns 2000; Luo, 2002a) but the formulae for the three key developments, CMC(l:c), CIE94 and CIEDE2000, are given. The CMC equation, developed in the early 1980s, was a key development in colour science and became a standard in certain countries and industries (Clarke *et al.*, 1984). It

---

Computational Colour Science Using MATLAB. By Stephen Westland and Caterina Ripamonti.
© 2004 John Wiley & Sons, Ltd: ISBN 0 470 84562 7

was never adopted by the CIE as a standard, however, and by the early 1990s there was some concern that the formula might be overly complex and that its predictions might be poor in certain areas of colour space. The CIE recommended the CIE94 equation (Berns, 1993) for use before a concerted effort was made to develop a new formula. CIEDE2000 (Luo *et al.*, 2001) was developed following collaboration between scientists working in several countries and has now been adopted as a CIE recommendation for the prediction of small colour differences.

5.2 CIELAB and CIELUV colour space

Between 1940 and 1976 a great number of colour spaces, transformations of *XYZ*, were proposed as uniform colour spaces. Some of these, such as HunterLab and ANLAB, were quite successful but in 1976 the CIE agreed upon two transformations that led to CIELAB and CIELUV.

The formulae for computing CIELAB coordinates are given in Equations (5.1):

$$L^* = 116(Y/Y_n)^{1/3} - 16, \quad \text{if } Y/Y_n > 0.008856,$$
$$L^* = 903.3(Y/Y_n), \qquad \text{if } Y/Y_n \leqslant 0.008856,$$

$$(5.1)$$

$$a^* = 500[\text{f}(X/X_n) - \text{f}(Y/Y_n)],$$
$$b^* = 200[\text{f}(Y/Y_n) - \text{f}(Z/Z_n)],$$

where

$$\text{f}(I) = (I)^{1/3}, \qquad \text{if } I > 0.008856,$$
$$\text{f}(I) = 7.787(I) + 16/116, \quad \text{if } I \leqslant 0.008856,$$

and where X_n, Y_n and Z_n are the tristimulus values of a specified white object colour. For surface colours the values of X_n, Y_n and Z_n usually are computed for the perfect reflecting diffuser and are therefore equivalent to the illuminant itself. Since white surfaces tend to look chromatically neutral under an illumination to which the visual system is adapted the values of X_n, Y_n and Z_n sometimes are referred to as the neutral point. The axes L^*, a^* and b^* form a rectangular or Cartesian coordinate space where L^* represents lightness, a^* represents redness-greenness and b^* represents yellowness-blueness. Sometimes it is useful to represent colour stimuli in a cylindrical space and for these purposes it is possible to compute the polar coordinates C_{ab}^* and h_{ab} as shown in Equations (5.2) and (5.3),

$$C_{ab}^* = (a^{*2} + b^{*2})^{1/2}, \qquad (5.2)$$

$$h_{ab} = \tan^{-1}(b^*/a^*)(180/\pi), \tag{5.3}$$

where the term $180/\pi$ is necessary to convert the output of the inverse tan function from radians to degrees. The polar coordinates are useful since the differences in the chroma term C^*_{ab} can be correlated with differences in perceived colourfulness, and differences in the hue term h_{ab} can be correlated with differences in perceived hue. Equations (5.2) and (5.3) can easily be inverted (Green, 2002a),

$$a^* = C^* \cos(h_{ab}\pi/180), \tag{5.4}$$

$$b^* = C^* \sin(h_{ab}\pi/180). \tag{5.5}$$

If the tristimulus values of the neutral are known, then it is possible to invert Equations (5.1),

$$
\begin{aligned}
Y &= Y_n \mathrm{f}(Y/Y_n)^3, & &\text{if } \mathrm{f}(Y/Y_n) > (0.008856)^{1/3}, \\
Y &= Y_n(\mathrm{f}(Y/Y_n) - 16/116)/7.787), & &\text{if } \mathrm{f}(Y/Y_n) \leqslant (0.008856)^{1/3}, \\
X &= X_n\, \mathrm{f}(X/X_n)^3, & &\text{if } \mathrm{f}(X/X_n) > (0.008856)^{1/3}, \\
X &= X_n(\mathrm{f}(X/X_n) - 16/116)/7.787), & &\text{if } \mathrm{f}(X/X_n) \leqslant (0.008856)^{1/3}, \\
Z &= Z_n \mathrm{f}(Z/Z_n)^3, & &\text{if } \mathrm{f}(Z/Z_n) > (0.008856)^{1/3}, \\
Z &= Z_n(\mathrm{f}(Z/Z_n) - 16/116)/7.787), & &\text{if } \mathrm{f}(Z/Z_n) \leqslant (0.008856)^{1/3},
\end{aligned}
\tag{5.6}
$$

where

$$\mathrm{f}(Y/Y_n) = (L^* + 16)/116,$$
$$\mathrm{f}(X/X_n) = a^*/500 + \mathrm{f}(Y/Y_n)$$

and

$$\mathrm{f}(Z/Z_n) = \mathrm{f}(Y/Y_n) - b^*/200.$$

The formulae for computing CIELUV coordinates are given as Equations (5.7),

$$
\begin{aligned}
L^* &= 116(Y/Y_n)^{1/3} - 16, & &\text{if } Y/Y_n > 0.008856, \\
L^* &= 903.3(Y/Y_n), & &\text{if } Y/Y_n \leqslant 0.008856,
\end{aligned}
$$

$$
\begin{aligned}
u^* &= 13L^*(u' - u'_n), \\
v^* &= 13L^*(v' - v'_n),
\end{aligned}
\tag{5.7}
$$

where u' and v' are the coordinates of the so-called uniform chromaticity space, CIE 1976 UCS, which is a linear transform of the more usual xy chromaticity space,

$$u' = 4X/(X + 15Y + 3Z),$$
$$v' = 9Y/(X + 15Y + 3Z). \tag{5.8}$$

The subscript n in Equations (5.7) again refers to the neutral point. It is also possible to compute polar coordinates for CIELUV,

$$C^*_{uv} = (u^{*2} + v^{*2})^{1/2},$$
$$h_{uv} = \tan^{-1}(v^*/u^*)(180/\pi). \tag{5.9}$$

Whereas CIELAB was recommended for use with surface colours, CIELUV was recommended for use with self-luminous colours (surface colours are sometimes referred to as related colours since we rarely see a surface in isolation but rather as part of a scene). One of the reasons for this is that the CIELUV space retains a chromaticity diagram which is derived by plotting u' against v'. An approximately uniform chromaticity space is useful since the additive mixtures of two stimuli all lie on the straight line in chromaticity space between the points that represent the two stimuli. However, in the last couple of decades CIELAB has become almost exclusively used for colour specification and the vast majority of work on the prediction of colour difference and colour appearance has been based upon CIELAB. It has been noted that there seems no reason to use CIELUV over CIELAB (Fairchild, 1998).

5.3   CIELAB colour difference

The CIELAB space has become popular largely because of the associated colour-difference metric [Equation (5.10)] that is computed as the Euclidean distance between two points in CIELAB space,

$$\Delta E^*_{ab} = [(\Delta L^*)^2 + (\Delta a^*)^2 + (\Delta b^*)^2]^{1/2}, \tag{5.10}$$

where

$$\Delta L^* = L^*_T - L^*_S,$$
$$\Delta a^* = a^*_T - a^*_S,$$
$$\Delta b^* = b^*_T - b^*_S,$$

and the subscripts refer to the standard (S) and the trial (T). In industrial applications of colour difference it is common that one of the samples is a standard and the other is a sample or trial that is supposed to be a visual match to the standard.

   An idea of the size of ΔE^*_{ab} units can be gained by considering that the difference between a perfect white ($L^* = 100$, $a^* = b^* = 0$) and a perfect black

$(L* = a* = b* = 0)$ is 100 ΔE^*_{ab} units, whereas industrial tolerances usually are about 1.0 CIELAB units.

If we imagine ΔE^*_{ab} to be computed from polar coordinates, then we could write an equivalent equation in terms of ΔL^*, ΔC^*_{ab} and ΔH^*_{ab},

$$\Delta E^*_{ab} = [(\Delta L^*)^2 + (\Delta C^*_{ab})^2 + (\Delta H^*_{ab})^2]^{1/2}, \tag{5.11}$$

where ΔH^*_{ab} is the difference in hue that is both commensurate with the other variables of CIELAB colour difference and orthogonal to both ΔL^* and ΔC^*_{ab}. Whereas the other terms are computed as simple differences (ΔC^* is simply the difference between C^* of the standard and the trial) ΔH^*_{ab} is defined by equating Equations (5.10) and (5.11) to yield the algebraic expression given as Equation (5.12) (Smith, 1997). Thus,

$$\Delta H^*_{ab} = [(\Delta E^*_{ab})^2 - (\Delta L^*)^2 - (\Delta C^*_{ab})^2]^{1/2} \tag{5.12}$$

or simply

$$\Delta H^*_{ab} = [(\Delta a^*)^2 + (\Delta b^*)^2 - (\Delta C^*_{ab})^2]^{1/2}.$$

Sometimes an alternative method is used to compute a hue difference as given by Equation (5.13),

$$\Delta H^*_{ab} = C^*_{ab}\Delta h_{ab}(\pi/180), \tag{5.13}$$

where the term $\pi/180$ converts the difference in hue angle Δh_{ab} into radians. However, Smith (1997) notes that this method is only applicable for small colour differences away from the achromatic axis and prefers Equation (5.12) which is more generally applicable.

Normally when a colour difference is computed between two samples one of the samples is regarded as the standard and the other as the trial or batch. The components of the colour difference therefore have a positive or negative sign and are computed as, for example, the chroma of the trial minus the chroma of the standard. Thus, if $\Delta C^*_{ab} > 0$, then the trial is stronger than the standard, whereas if $\Delta C^*_{ab} < 0$, then the trial is weaker than the standard. Similarly, the trial can be lighter or darker than the standard depending upon the sign of the ΔL^* component. However, the definition of the hue component of colour difference as in Equation (5.12) leads to some ambiguity in the sign of ΔH^*_{ab}. By convention, it is to be regarded as positive if it indicates that in terms of hue the trial is anticlockwise from the standard and negative if it is clockwise.

The signs of the colour-difference components are most useful in determining colour-difference descriptors between a trial and a standard. Whereas the determination that the trial is either stronger or weaker and lighter or darker than the standard derives simply from the sign of ΔL^* or ΔC^*_{ab}, the determination of hue difference descriptors is more complicated. The CIE

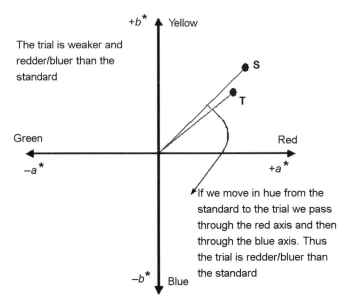

The trial is weaker and redder/bluer than the standard

If we move in hue from the standard to the trial we pass through the red axis and then through the blue axis. Thus the trial is redder/bluer than the standard

Figure 5.1   Example calculation of hue difference descriptors for a standard (*S*) and trial (*T*) sample in CIELAB colour space

recommends that two hue descriptors be assigned to any pair of samples. An illustration is given by the standard and trial samples represented in Figure 5.1.

In Figure 5.1 the chroma of the trial is smaller than that of the standard and therefore we can deduce that the trial is weaker than the standard. To derive a hue difference descriptor we move from the standard to the trial in the hue circle and note the first two axes that are crossed. In the example illustrated by Figure 5.1, we move in a clockwise direction (as denoted by the arrow) and pass through the red axis and then the blue axis. It is usual then to describe the trial as being redder/bluer than the standard. Note that the value of ΔH^*_{ab} would be assigned a negative sign since the trial is clockwise from the standard. Why should the trial be described as being redder/bluer? Would it not be simpler to use the closest axis and denote the trial as being redder? The answer is that the correct choice of hue descriptor is difficult to predict in advance without knowledge of the colour appearance of the samples (the issue of colour appearance will be discussed in more detail in Chapter 6). If, in Figure 5.1, the two samples appear yellow, then it would be reasonable to describe the trial as being redder than the standard. However, if the colour appearance of the two samples is essentially red, then it is not informative in terms of hue difference to describe the trial as being redder than the standard; rather, in this case we would say that the trial is bluer. For two samples in the first quadrant (that is, *a** and *b** are both positive) it is possible for the two samples to appear yellow (which is likely if the samples are close to the yellow axis) or red (which is likely if the samples are closer to the red axis) and therefore most computer programs that compute colour difference report both possible hue-difference terms (redder/bluer in the example).

Figure 5.1 emphasizes why the polar coordinates C_{ab}^* and h_{ab} usually are preferred to the cartesian coordinates a^* and b^*. The fact that the trial has a smaller a^* value than the standard ($\Delta a^* < 0$) could be misinterpreted as indicating that the standard is redder than the trial and yet the opposite is true in terms of hue; the trial is redder/bluer than the standard. The possible error occurs because the dimensions of human colour perception are brightness, colourfulness and hue and these correlate with the polar coordinates lightness, chroma and hue. It can be misleading to consider differences in a^* and b^* in isolation since these confound differences in chroma and hue.

5.4 Optimized colour-difference formulae

5.4.1 CMC(*l*:*c*)

In the late 1970s a number of different formulae were being used by practitioners, one of which was known as the JPC79 formula (the name derives from J. & P. Coats whose laboratories developed the formula). The JPC79 formula was effective but was known to be deficient in some areas (Smith, 1997) and a revised version of the formula was published in 1984 by members of the Colour Measurement Committee of the Society of Dyers and Colourists (Clarke *et al.*, 1984). This revised formula became known as CMC(*l*:*c*) which, like most modern optimized colour-difference formulae, is based upon the CIELAB colour-difference components ΔL^*, ΔC_{ab}^* and ΔH_{ab}^*,

$$\Delta E_{\mathrm{CMC}(l:c)} = [(\Delta L^*/(lS_L))^2 + (\Delta C_{ab}^*/(cS_C))^2 + (\Delta H_{ab}^*/S_H)^2]^{1/2}, \qquad (5.14)$$

where $S_L = 0.040975 L_S^*/(1 + 0.01765 L_S^*)$,   if $L_S^* \geqslant 16$,
but $S_L = 0.511$,   if $L_S^* < 16$,

and

$S_C = 0.638 + 0.0638 C_{ab,S}^*/(1 + 0.0131 C_{ab,S}^*)$,
$S_H = S_C(TF + 1 - F)$.

The terms T and F are given by

$$F = [(C_{ab,S}^*)^4/((C_{ab,S})^4 + 1900)]^{1/2}$$

and

$T = 0.36 + |0.4\cos(h_{ab,S} + 35)|$,   if $h_{ab,S} \leqslant 164$ or $h_{ab,S} \geqslant 345$,
$T = 0.56 + |0.2\cos(h_{ab,S} + 168)|$, if $164 < h_{ab,S} < 345$.

The subscript S, as in C_S, denotes that the terms S_L, S_C, S_H, F and T are computed using the CIELAB lightness, chroma and hue angle (in degrees) of the standard. The terms S_L, S_C and S_H define the lengths of the semi-axes of the tolerance ellipsoid at the position of the standard in CIELAB space in each of the three directions (S_L for lightness, S_C for chroma and S_H for hue). The ellipsoids were fitted to visual tolerances determined from psychophysical experiments and the semi-axes in the CMC(l:c) formula are used to effectively convert these ellipsoids into spheres at each point in CIELAB space. The parametric terms l and c constitute an important feature of the formula. These parameters allow the relative tolerances of the lightness and chroma components to be modified. For the textile industry it was recommended that $l = c = 1$ for perceptibility decisions, whereas for acceptability decisions it was recommended that $l = 2$ with $c = 1$. The reason for this difference is that it is considered that, in terms of acceptability, differences in lightness should be weighted to be half as important as differences in either chroma or hue. The CMC(l:c) formula has been widely used in a number of industries and was adopted, for example, as a British Standard (BS 6923) and an AATCC test method (AATCC 173). However, it was never adopted as a CIE standard.

5.4.2   CIE94

Berns (2000) and others argued that the complexity of the CMC equation and the use of large numbers of significant figures in its definition suggest a degree of precision that cannot be supported on statistical grounds. Detailed analyses of large sets of psychophysical data suggested that simple S_L, S_C and S_H weighting functions would be sufficient and this led to the publication of a new formula known as CIE94 (Berns, 1993). The CIE94 formula is given by

$$\Delta E_{94}^* = [(\Delta L^*/(k_L S_L))^2 + (\Delta C_{ab}^*/(k_C S_C))^2 + (\Delta H_{ab}^*/k_C S_H)^2]^{1/2}, \qquad (5.15)$$

where

$$S_L = 1,$$
$$S_C = 1 + 0.045 C_{ab,S}^*,$$
$$S_H = 1 + 0.015 C_{ab,S}^*.$$

The parametric variables k_L, k_C and k_H are all set to unity and the values of S_C and S_H are computed using the CIELAB values of the standard. When neither sample can logically be deemed a standard, the geometric mean of the two samples should be used.

5.4.3   CIEDE2000

The CIE have recently recommended for trial the CIEDE2000 colour-difference formula for the evaluation of small colour differences (Luo *et al.*, 2001). Note that CIELAB ΔE^*_{ab} is still the current CIE recommendation for the evaluation of large colour differences ($\Delta E^*_{ab} > 5$).

The CIEDE2000 formula was agreed by a technical committee within Division 1 of the CIE (2001) and includes not only lightness, chroma and hue weighting functions, but also an interactive term between the chroma and hue differences for improving the performance for blue colours and a scaling factor for the CIELAB $a*$ scale for improving the performance for colours close to the achromatic axis. The new formula is given by Equation (5.16),

$$\Delta E_{00} = [(\Delta L'/(k_L S_L))^2 + (\Delta C'/(k_C S_C))^2 + (\Delta H'/(k_H S_H))^2$$
$$+ R_T(\Delta C'/(k_C S_C))(\Delta H'/(k_H S_H))]^{1/2}, \tag{5.16}$$

where

$$S_L = 1 + [0.015(L' - 50)^2]/[20 + (L' - 50)^2]^{1/2},$$
$$S_C = 1 + 0.045C',$$
$$S_H = 1 + 0.015C'T.$$

The terms $\Delta L'$, $\Delta C'$ and $\Delta H'$ are given by

$$\Delta L' = L'_T - L'_S,$$
$$\Delta C' = C'_T - C'_S,$$
$$\Delta H' = 2(C'_T C'_S)^{1/2} \sin(\Delta h'/2),$$

where the subscripts S and T refer to the standard and trial, respectively, and where

$$\Delta h' = h'_T - h'_S,$$
$$L' = L^*,$$
$$a' = (1 + G)a^*,$$
$$b' = b^*,$$
$$C' = (a'^2 + b'^2)^{1/2},$$

and

$$h' = \tan^{-1}(b'/a').$$

The G and T terms are computed using

$$G = 0.5 - 0.5(C^{*7}_{ab}/(C^{*7}_{ab} + 25^7))^{1/2}$$

and

$$T = 1 - 0.17\cos(h' - 30) + 0.24\cos(2h') + 0.32\cos(3h' + 6) - 0.20\cos(4h' - 63).$$

Finally, the rotation term R_T is given by

$$R_T = -\sin(2\Delta\theta)R_C,$$

where $R_C = 2(C'^7/(C'^7 + 25^7))^{1/2}$ and $\Delta\theta = 30\exp\{-[(h' - 275)/25]^2\}$.

Note that the arithmetic mean of the CIELAB values of the standard and trial are used to compute the values of the terms such as S_L. The CIEDE2000 formula has been shown to outperform the CMC and CIE94 formulae by a large margin (Luo *et al.*, 2001).

Interestingly, the lightness component of the formula is very different from those in earlier formulae. For example, the value of the function in the CMC formula increases markedly as L^* increases, implying that for equal differences in L^* the visual difference should be largest in the low L^* region. The lightness correction in the CIE94 formulae, on the other hand, implied that the CIELAB L^* scale was correct, so that equal differences in L^* would yield equal visual differences no matter what the value of L^*. The S_L formula in CIEDE2000, however, was based upon new data (Heptinstall, 1999; Chou *et al.*, 2001) so that S_L increases with L^* only for $L^* > 50$; for lower values of L^* the value of S_L decreases as L^* increases. It is still not at all clear why the new data upon which CIEDE2000 was based should have been so different from the data upon which the earlier formulae were based. Nevertheless, the evidence for CIEDE2000 is convincing and there is strong confidence that the new formula is reliable (Cui *et al.*, 2001; Luo, 2002a).

5.5 Implementations and examples

5.5.1 Computing CIELAB and CIELUV coordinates

The function *xyz2lab* computes CIELAB $L^*a^*b^*$ coordinates from tristimulus values. A typical call would be

```
[lab] = xyz2lab(xyz,'d65_64');
```

where the variable **xyz** is a 3×1 vector of tristimulus values and **lab** returns the CIELAB L^*, a^* and b^* values. The white points are taken from Table 5 of the ASTM standard (ASTM, 2001) which are reproduced in Table 4.2. Note that the ASTM standard specifies that the white points listed at the bottom of each of the tables in the standard should be used for the values of X_n, Y_n and Z_n during computations where the neutral point is required. The listed white points sometimes differ from the check sums for each of the tables because the tabulated data are rounded to three decimal places. The quoted white points are more accurate than the sums of the rounded data in the columns. However, a consequence of the ASTM recommendation is that the CIELAB values for a sample with unit reflectance at every wavelength may not exactly satisfy $L^* = 100$ and $a^* = b^* = 0$. For example, the XYZ values using ASTM Table 5.19 (illuminant D65 and 1964 observer) for a perfectly reflecting sample are [94.809 100.00 107.307] but the ASTM white point for that illuminant/ observer is [94.811 100.00 107.304]. Consequently, the ratios X/X_n, Y/Y_n and Z/Z_n for a perfectly reflecting sample are not exactly unity and the CIELAB $L^*a^*b^*$ values returned from *xyz2lab* are [100.0000 −0.0035 −0.0019].

Box 5: *xyz2lab.m*

```
function [lab] = xyz2lab(xyz,obs)

% function [lab] = xyz2lab(xyz,obs)
% computes CIELAB LAB values from XYZ tristimulus values
% requires the illuminant/observer obs to define white
% point
% see function r2xyz for valid values for obs

if strcmp('a_64',obs)
  white=[111.144 100.00 35.200];
elseif strcmp('a_31', obs)
  white=[109.074 100.00 35.585];
elseif strcmp('c_64', obs)
  white=[97.285 100.00 116.145];
elseif strcmp('c_31', obs)
  white=[98.074 100.00 118.232];
elseif strcmp('d50_64', obs)
  white=[96.720 100.00 81.427];
elseif strcmp('d_50', obs)
  white=[96.422 100.00 82.521];
elseif strcmp('d55_64', obs)
  white=[95.799 100.00 90.926];
elseif strcmp('d55_31', obs)
  white=[95.682 100.00 92.149];
elseif strcmp('d65_64', obs)
  white=[94.811 100.00 107.304];
elseif strcmp('d65_31', obs)
  white=[95.047 100.00 108.883];
elseif strcmp('d75_64', obs)
```

```
  white=[94.416 100.00 120.641];
elseif strcmp('d75_31',obs)
  white=[94.072 100.00 122.638];
elseif strcmp('f2_64',obs)
  white=[103.279 100.00 69.027];
elseif strcmp('f2_31',obs)
  white=[99.186 100.00 67.393];
elseif strcmp('f7_64',obs)
  white=[95.792 100.00 107.686];
elseif strcmp('f7_31',obs)
  white=[95.041 100.00 108.747];
elseif strcmp('f9_64',obs)
  white=[103.863 100.00 65.607];
elseif strcmp('f9_31',obs)
  white=[100.962 100.00 64.350];
else
  disp('unknown option obs');
  disp('use d65_64 for D65 and 1964 observer'); return;
end

dim = size(xyz);
if (dim(1) == 1) | (dim(2) == 1)
  xyz = xyz(:)'; % force to be a row matrix
else
  disp('xyz must be a row matrix');
  return;
end

if (xyz(2)/white(2) > 0.008856)
  lab(1) = 116*(xyz(2)/white(2))^(1/3) - 16;
else
  lab(1) = 903.3*(xyz(2)/white(2));
end

for i=1:3
  if (xyz(i)/white(i) > 0.008856)
    fx(i) = (xyz(i)/white(i))^(1/3);
  else
    fx(i) = 7.787*(xyz(i)/white(i)) + 16/116;
  end
end
```

```
lab(2) = 500*(fx(1)-fx(2));
lab(3) = 200*(fx(2)-fx(3));
```

To fully utilize the advantages of matrix algebra in MATLAB the function *xyz2lab* could be written to accept a $3 \times n$ matrix that would contain the tristimulus values of n samples (where n is any positive integer) and would return a $3 \times n$ matrix of CIELAB values. However, in this book the functions generally have not been written in this way and if transformations are required for n samples, then the functions must be called n times using a programming loop. It is relatively straightforward to invert the CIELAB equations if the white point is known. A function called *lab2xyz* has been provided for this purpose and has the following typical function call:

```
[xyz] = lab2xyz(lab,'d65_64');
```

where the variable **lab** is a 3×1 vector of CIELAB L^*, a^* and b^* values.

Box 6: *lab2xyz.m*

```
function [xyz] = lab2xyz(lab,obs)

% function [xyz] = lab2xyz(lab,obs)
% computes XYZ tristimulus values from CIELAB LAB values
% requires the illuminant/observer obs to define white
% point
% see function r2xyz for valid values for obs

if strcmp('a_64',obs)
  white=[111.144 100.00 35.200];
elseif strcmp('a_31', obs)
  white=[109.074 100.00 35.585];
elseif strcmp('c_64', obs)
  white=[97.285 100.00 116.145];
elseif strcmp('c_31', obs)
  white=[98.074 100.00 118.232];
elseif strcmp('d50_64', obs)
  white=[96.720 100.00 81.427];
elseif strcmp('d_50', obs)
  white=[96.422 100.00 82.521];
```

```
elseif strcmp('d55_64', obs)
  white=[95.799 100.00 90.926];
elseif strcmp('d55_31', obs)
  white=[95.682 100.00 92.149];
elseif strcmp('d65_64', obs)
  white=[94.811 100.00 107.304];
elseif strcmp('d65_31', obs)
  white=[95.047 100.00 108.883];
elseif strcmp('d75_64', obs)
  white=[94.416 100.00 120.641];
elseif strcmp('d75_31', obs)
  white=[94.072 100.00 122.638];
elseif strcmp('f2_64', obs)
  white=[103.279 100.00 69.027];
elseif strcmp('f2_31', obs)
  white=[99.186 100.00 67.393];
elseif strcmp('f7_64', obs)
  white=[95.792 100.00 107.686];
elseif strcmp('f7_31', obs)
  white=[95.041 100.00 108.747];
elseif strcmp('f9_64', obs)
  white=[103.863 100.00 65.607];
elseif strcmp('f9_31', obs)
  white=[100.962 100.00 64.350];
else
  disp('unknown option obs');
  disp('use d65_64 for D65 and 1964 observer'); return;
end

dim = size(lab);
if (dim(1) == 1) | (dim(2) == 1)
  lab = lab(:)'; % force to be a row matrix
else
  disp('lab must be a row matrix');
  return;
end

% compute Y
if (((lab(1)+16)/116)^3 > 0.008856)
  xyz(2) = white(2)*((lab(1)+16)/116)^3;
else
  xyz(2) = white(2)*lab(1)/903.3;
```

```
end
% compute fy for use later
fy = xyz(2)/white(2);
if (fy > 0.008856)
   fy = fy^(1/3);
else
   fy = 7.787*fy + 16/116;
end

% compute X
if ((lab(2)/500 + fy)^3 > 0.008856)
   xyz(1) = white(1)*(lab(2)/500 + fy)^3;
else
   xyz(1) = white(1)*((lab(2)/500 + fy) - 16/116)/7.787;
end

% compute Z
if ((fy - lab(3)/200)^3 > 0.008856)
   xyz(3) = white(3)*(fy - lab(3)/200)^3;
else
   xyz(3) = white(3)*((fy - lab(3)/200) - 16/116)/7.787;
end
```

The function *xyz2luv* computes CIELUV $L^*u^*v^*$ coordinates from tristimulus values. A typical call would be

```
[luv] = xyz2luv(xyz,'d65_64');
```

where **xyz** is a 3×1 column matrix of tristimulus values. The previous discussion about the selection of the white point in *xyz2lab* is equally valid in the case of *xyz2luv*. An alternative function call is also possible, however, so that the u' and v' values of the CIE 1976 UCS can also be obtained:

```
[luv, uprime, vprime] = xyz2luv(xyz,'d65_64');
```

In **MATLAB** whenever multiple outputs are returned from a function, typing the function on its own

```
xyz2luv(xyz,'d65_64')
```

will output only the first of these arguments (in this case the vector **luv**) and assign it to the variable **ans**.

Box 7: *xyz2luv.m*

```
function [luv,uprime,vprime] = xyz2luv(xyz,obs)

% function [luv,uprime,vprime] = xyz2luv(xyz,obs)
% computes CIELUV Luv values from XYZ tristimulus values
% uprime and vprime are the CIE 1976 UCS coordinates
% requires the illuminant/observer obs to define white
% point
% see function r2xyz for valid values for obs

if strcmp('a_64',obs)
  white=[111.144 100.00 35.200];
elseif strcmp('a_31', obs)
  white=[109.074 100.00 35.585];
elseif strcmp('c_64', obs)
  white=[97.285 100.00 116.145];
elseif strcmp('c_31', obs)
  white=[98.074 100.00 118.232];
elseif strcmp('d50_64', obs)
  white=[96.720 100.00 81.427];
elseif strcmp('d_50', obs)
  white=[96.422 100.00 82.521];
elseif strcmp('d55_64', obs)
  white=[95.799 100.00 90.926];
elseif strcmp('d55_31', obs)
  white=[95.682 100.00 92.149];
elseif strcmp('d65_64', obs)
  white=[94.811 100.00 107.304];
elseif strcmp('d65_31', obs)
  white=[95.047 100.00 108.883];
elseif strcmp('d75_64', obs)
  white=[94.416 100.00 120.641];
elseif strcmp('d75_31', obs)
  white=[94.072 100.00 122.638];
elseif strcmp('f2_64', obs)
  white=[103.279 100.00 69.027];
elseif strcmp('f2_31', obs)
  white=[99.186 100.00 67.393];
elseif strcmp('f7_64', obs)
  white=[95.792 100.00 107.686];
```

```
elseif strcmp('f7_31',obs)
  white=[95.041 100.00 108.747];
elseif strcmp('f9_64',obs)
white=[103.863 100.00 65.607];
elseif strcmp('f9_31',obs)
  white=[100.962 100.00 64.350];
else
  disp('unknown option obs');
  disp('use d65_64 for D65 and 1964 observer'); return;
end

dim = size(xyz);
if (dim(1) == 1) | (dim(2) == 1)
  xyz = xyz(:)'; % force to be a row matrix
else
  disp('xyz must be a row matrix');
  return;
end

% compute u' v' for sample
uprime = 4*xyz(1)/(xyz(1) + 15*xyz(2) + 3*xyz(3));
vprime = 9*xyz(2)/(xyz(1) + 15*xyz(2) + 3*xyz(3));
% compute u' v' for white
uprimew = 4*white(1)/(white(1) + 15*white(2) + ...
3*white(3));
vprimew = 9*white(2)/(white(1) + 15*white(2) + ...
3*white(3));

if (xyz(2)/white(2) > 0.008856)
  luv(1) = 116*(xyz(2)/white(2))^(1/3) - 16;
else
  luv(1) = 903.3*(xyz(2)/white(2));
end

luv(2) = 13*luv(1)*(uprime - uprimew);
luv(3) = 13*luv(1)*(vprime - vprimew);
```

A single function *car2pol* has been provided to compute polar coordinates from cartesian coordinates. If the input to this function is a^* and b^*, then the output is C_{ab}^* and h_{ab}, whereas if the input is u^* and v^*, then the output is C_{uv}^* and h_{uv}.

Box 8: *car2pol.m*

```
function [c,h] = car2pol(ab)

% function [c,h] = cartopol(ab)
% converts a*b* or u*v* into the polar coordinates
% of Chroma C and Hue H
% ab must be a row or column matrix 2 by 1 or 1 by 2
% see also pol2car

dim = size(ab);
if (dim(1) == 1) | (dim(2) == 1)
  ab = ab(:)'; % force to be a row matrix
else
  disp('ab must be a row matrix');
  return;
end
if (dim(2) ~= 2)
  disp('ab must be 2 by 1 or 1 by 2');
  return;
end

% compute the distance from the centre
c = sqrt(ab(1)*ab(1) + ab(2)*ab(2));

% compute the angular term
if (ab(1) == 0) & (ab(2) > 0)
  h = 90;
elseif (ab(1) == 0) & (ab(2) < 0)
  h = 270;
elseif (ab(1) < 0) & (ab(2) == 0)
  h = 180;
elseif (ab(1) > 0) & (ab(2) == 0)
  h = 0;
elseif (ab(1) == 0) & (ab(2) == 0)
  h = 0;
else
  h = atan(abs(ab(2))/abs(ab(1)));
  h = 180*h/pi; % convert from radians to degrees
  if ((ab(1) > 0) & (ab(2) > 0))
    h = h; % first quadrant
```

```
  elseif ((ab(1) < 0) & (ab(2) > 0))
    h = 180 - h; % second quadrant
  elseif ((ab(1) < 0) & (ab(2) < 0))
    h = 180 + h; % third quadrant
  else
    h = 360 - h; % fourth quadrant.
  end
end
```

The syntax for the function call is

```
[c,h] = car2pol(ab)
```

where **ab** is a 2×1 or 1×2 matrix and c and h are both 1×1 matrices whose entries are the polar coordinates. Note that the code in *car2pol* could be shortened considerably by the use of the *atan2* command. Whereas the *atan* MATLAB function returns the arctangent of the input element and requires that the quadrant be determined from the polarities of the cartesian coordinates, the *atan2* command returns the four-quadrant arctangent directly.

The function *pol2car* is provided to return polar coordinates to cartesian coordinates. The format for the function call is

```
[a,b] = pol2car(ch)
```

where **ch** is a 2×1 or 1×2 matrix containing the distance and angular terms and **a** and **b** are both 1×1 matrices whose entries are the horizontal and vertical components of the cartesian space.

Box 9: *pol2car.m*

```
function [a,b] = pol2car(ch)

% function [a,b] = pol2car(ch)
% converts the polar coordinates
% of Chroma C and Hue H
% ch must be a row or column matrix 2 by 1 or 1 by 2
% see also car2pol

dim = size(ch);
if (dim(1) == 1) | (dim(2) == 1)
  ch = ch(:)'; % force to be a row matrix
```

```
  else
    disp('ch must be a row matrix');
    return;
  end
  if (dim(2) ~= 2)
    disp('ch must be 2 by 1 or 1 by 2');
    return;
  end

  C = ch(1);
  H = ch(2);

  fx = tan(H*pi/180);

  a = sqrt(C*C/(1 + fx*fx));

  b = a*fx;

  if (H < 90.0)
    % first quadrant
    a = abs(a);
    b = abs(b);
  elseif (H < 180)
    % second quadrant
    a = -abs(a);
    b = abs(b);
  elseif (H < 270)
    % third quadrant
    a = -abs(a);
    b = -abs(b);
  else
    % fourth quadrant
    a = abs(a);
    b = -abs(b);
  end
```

5.5.2 Computing colour difference

The function *cielabde* computes the CIELAB colour difference [Equation (5.10)]
from two L^*, a^*, b^* triplets. This function also returns the component deltas,

ΔL^*, ΔC_{ab}^*, and ΔH_{ab}^*, in addition to the overall colour difference. A typical call would be

```
[de, dl, dc, dh] = cielabde(lab1, lab2);
```

where **lab1** and **lab2** are 3×1 column matrices containing the L^*, a^* and b^* values of the standard and trial, respectively. A shorter function call is of course also valid; thus

```
[de] = cielabde(lab1, lab2);
```

can be used if the individual component differences are not required.

Box 10: *cielabde.m*

```
function [de,dl,dc,dh] = cielabde(lab1,lab2)

% function [de,dl,dc,dh] = cielabde(lab1,lab2)
% computes colour difference from CIELAB values
% using CIELAB formula
% lab1 and lab2 must be 3 by 1 or 1 by 3 matrices
% and contain L*, a* and b* values
% see also cmcde, cie94de, and cie00de

dim = size(lab1);
if (dim(1) == 1) | (dim(2) == 1)
  lab1 = lab1(:)'; % force to be a row matrix
else
  disp('lab1 must be a row matrix');
  return;
end
if (dim(2) ~= 3)
  disp('lab1 must be 3 by 1 or 1 by 3');
  return;
end

dim = size(lab2);
if (dim(1) == 1) | (dim(2) == 1)
  lab2 = lab2(:)'; % force to be a row matrix
else
```

```
   disp('lab2 must be a row matrix');
   return;
end
if (dim(2) ~= 3)
   disp('lab2 must be 3 by 1 or 1 by 3');
   return;
end

dl = lab2(1)-lab1(1);

dc = sqrt(lab2(2)^2 + lab2(3)^2) - sqrt(lab1(2)^2 + ...
lab1(3)^2);
dh = sqrt((lab2(2)-lab1(2))^2 + (lab2(3)-lab1(3))^2 - ...
dc^2);

% get the polarity of the dh term
dh = dh*dhpolarity(lab1,lab2);

de = sqrt(dl^2 + dc^2 + dh^2);
```

The function *cielabde* uses Equation (5.12) to compute the hue difference. As previously mentioned this returns the magnitude of the hue difference but the sign is indeterminate. A separate function has therefore been written called *dhpolarity* which returns $+1$ if the trial is anti-clockwise from the standard and -1 if the trial is clockwise from the standard. The format for this function is

```
   [p] = dhpolarity(lab1, lab2);
```

where **lab1** and **lab2** are the 3×1 column matrices containing the L^*, a^* and b^* values of the standard and trial, respectively.

Box 11: *dhpolarity.m*

```
function [p] = dhpolarity(lab1,lab2)

% function [p] = dhpolarity(lab1,lab2)
% computes polarity of hue difference
```

```
% lab1 and lab2 must be 3 by 1 or 1 by 3 matrices
% and contain L*, a* and b* values
% p is +1 if the hue of the trial (lab2) is anticlockwise
% from the standard (lab1) and -1 otherwise

[c1, h1] = car2pol([lab1(2) lab1(3)]);
[c2, h2] = car2pol([lab2(2) lab2(3)]);
p = (h2-h1);
if (p==0)
  p = 1;
else
  if (p>180)
    p = p - 360;
  end
  p = p/abs(p);
end
```

The function *cmcde* operates in a similar way to *cielabde* but computes the CMC(*l*:*c*) colour difference. A typical function call would be

```
[de, dl, dc, dh] = cmcde(lab1, lab2, paral, parac);
```

where the variables **paral** and **parac** represent the parametric values *l* and *c*. Default values of 1 are used for both *l* and *c* if the number of arguments to the function is less than four. Note that the component delta values that are also returned are the CMC delta values rather than the CIELAB delta values. Note also that the MATLAB trigonometric functions expect input in radians and therefore the hue angle in degree must be multiplied by the factor $\pi/180$ when computing the parameter *T* in Equation (5.14). The dimensions of the CMC tolerance ellipsoids are computed based upon the CIELAB values of the first of the triplets **lab1** which is assumed to be the standard.

Box 12: *cmcde.m*

```
function [de,dl,dc,dh] = cmcde(lab1,lab2,paral,parac)

% function [de,dl,dc,dh] = cmcde(lab1,lab2,paral,parac)
% computes colour difference from CIELAB values
```

```
% using CMC(1:c) formula
% lab1 and lab2 must be 3 by 1 or 1 by 3 matrices
% and contain L*, a* and b* values
% The dl, dc and dh components are CMC deltas
% The defaults for paral and parac are 1
% see also cielabde, cie94de, and cie00de

dim = size(lab1);
if (dim(1) == 1) | (dim(2) == 1)
  lab1 = lab1(:)'; % force to be a row matrix
else
  disp('lab1 must be a row matrix');
  return;
end
if (dim(2) ~= 3)
  disp('lab1 must be 3 by 1 or 1 by 3');
  return;
end

dim = size(lab2);
if (dim(1) == 1)  (dim(2) == 1)
  lab2 = lab2(:)'; % force to be a row matrix
else
  disp('lab2 must be a row matrix');
  return;
end

if (dim(2) ~= 3)
  disp('lab2 must be 3 by 1 or 1 by 3');
  return;
end

if (nargin<4)
  disp('using default values of 1:c')
  paral=1; parac=1;
end

% first compute the CIELAB deltas
dl = lab2(1)-lab1(1);
dc = sqrt(lab2(2)^2 + lab2(3)^2) - sqrt(lab1(2)^2 +...
lab1(3)^2);
dh = sqrt((lab2(2)-lab1(2))^2 + (lab2(3)-lab1(3))^2 -...
```

```
dc^2);

% get the polarity of the dh term
dh = dh*dhpolarity(lab1,lab2);

% now compute the CMC weights
if (lab1(1)<16)
  Lweight = 0.511;
else
  Lweight = (0.040975*lab1(1))/(1 + 0.01765*lab1(1));
end
[c,h] = car2pol([lab1(2) lab1(3)]);
% require C*ab and H*ab of standard
Cweight = 0.638 + (0.0638*c)/(1 + 0.0131*c);
if (164 < h & h < 345)
  k1 = 0.56; k2 = 0.20; k3 = 168;
else
  k1 = 0.36; k2 = 0.40; k3 = 35;
end
T = k1 + abs(k2*cos((h + k3)*pi/180));

F = sqrt((c^4)/(c^4 + 1900));
Hweight = Cweight*(T*F + 1 - F);

dl = dl/(Lweight*paral);
dc = dc/(Cweight*parac);
dh = dh/Hweight;

de = sqrt(dl^2 + dc^2 + dh^2);
```

The function *cie94de* computes the CIE94 equation and operates in a similar manner to *cielabde* and *cmcde* with the following format:

```
[de, dl, dc, dh] = cie94de(lab1, lab2)
```

Box 13: *cie94de.m*

```
function [de,dl,dc,dh] = cie94de(lab1,lab2)

% function [de,dl,dc,dh] = cie94de(lab1,lab2)
```

```
% computes colour difference from CIELAB values
% using the CIE94 formula
% lab1 and lab2 must be 3 by 1 or 1 by 3 matrices
% and contain L*, a* and b* values
% The dl, dc and dh components are CIE94 deltas
% see also cielabde, cmcde, and cie00de

dim = size(lab1);
if (dim(1) == 1) | (dim(2) == 1)
  lab1 = lab1(:)'; % force to be a row matrix
else
  disp('lab1 must be a row matrix');
  return;
end
if (dim(2) ~= 3)
  disp('lab1 must be 3 by 1 or 1 by 3');
  return;
end

dim = size(lab2);
if (dim(1) == 1) | (dim(2) == 1)
  lab2 = lab2(:)'; % force to be a row matrix
else
  disp('lab2 must be a row matrix');
  return;
end
if (dim(2) ~= 3)
  disp('lab2 must be 3 by 1 or 1 by 3');
  return;
end

dl = lab2(1)-lab1(1);
dc = sqrt(lab2(2)^2 + lab2(3)^2) - sqrt(lab1(2)^2 + ...
lab1(3)^2);
dh = sqrt((lab2(2)-lab1(2))^2 + (lab2(3)-lab1(3))^2 - ...
dc^2);

% get the polarity of the dh term
dh = dh*dhpolarity(lab1,lab2);

% need to compute the weights
Lweight = 1.0;
```

```
[c,h] = car2pol([lab1(2) lab1(3)]);
% require C*ab and H*ab of standard
Cweight = 1.0 + 0.045*c;
Hweight = 1.0 + 0.015*c;

dl = dl/Lweight;
dc = dc/Cweight;
dh = dh/Hweight;
de = sqrt(dl^2 + dc^2 + dh^2);
```

Finally, the function *cie00de* implements the CIEDE2000 colour-difference metric:

```
[de, dl, dc, dh] = cie00de(lab1, lab2, paral, parac,
parah).
```

The CIEDE2000 formula allows for three parametric terms for lightness, chroma and hue weightings, respectively. The default values for these parameters are all set to unity. The implementation of the CIEDE2000 formula requires that the hue values of the standard and trial be averaged. But the arithmetic mean cannot simply be computed directly since this would give a mean hue of 185° for the two hues 20° and 350°, whereas the true average hue would be 5°. The approach taken here is to make use of the *pol2car* (Box 9) and *car2pol* (Box 8) functions. Polar representations are converted to cartesian representations so that the simple arithmetic means may be computed before returning to the polar representation to recover the average hue (see Box 13).

In this code an alternative procedure has been implemented to compute the hue difference:

$$\Delta H = 2(C_T C_S)^{1/2}\sin(\Delta h/2), \tag{5.17}$$

where Δh is the hue of the trial minus the hue of the standard. This method gives a relatively simple way to compute ΔH but a correction is still required to ensure the correct sign is always computed. The correction is to subtract 360 from Δh if $\Delta h > 180$.

Box 14: *cie00de.m*

```
function [de,dl,dc,dh] = cie00de(lab1,lab2,paral,
parac,parah)
```

```
% function [de,dl,dc,dh] = cie00de(lab1,lab2,paral,
% parac,parah)
% computes colour difference from CIELAB values
% using the CIEDE2000 formula
% lab1 and lab2 must be 3 by 1 or 1 by 3 matrices
% and contain L*, a* and b* values
% The dl, dc and dh components are CIEDE2000 deltas
% The defaults for paral, parac and parah are 1
% see also cielabde, cmcde, and cie94de

dim = size(lab1);
if (dim(1) == 1) | (dim(2) == 1)
  lab1 = lab1(:)'; % force to be a row matrix
else
  disp('lab1 must be a row matrix');
  return;
end
if (dim(2) ~= 3)
  disp('lab1 must be 3 by 1 or 1 by 3');
  return;
end

dim = size(lab2);
if (dim(1) == 1) | (dim(2) == 1)
  lab2 = lab2(:)'; % force to be a row matrix
else
  disp('lab2 must be a row matrix');
  return;
end

if (dim(2) ~= 3)
  disp('lab2 must be 3 by 1 or 1 by 3');
  return;
end

if (nargin<5)
  disp('using default values of parametric values')
  paral=1; parac=1; parah = 1;
end

% convert the cartesian a*b* to polar chroma and hue
```

```
[c1,h1] = car2pol([lab1(2) lab1(3)]);
[c2,h2] = car2pol([lab2(2) lab2(3)]);
meanC = (c2+c1)/2;

% compute the G factor using the arithmetic mean chroma
G = 0.5 - 0.5*sqrt((meanC^7)/(meanC^7 + 25^7));

% transform the a* values
lab1(2) = (1 + G)*lab1(2);
lab2(2) = (1 + G)*lab2(2);
% recompute the polar coordinates using the new a*
[c1,h1] = car2pol([lab1(2) lab1(3)]);
[c2,h2] = car2pol([lab2(2) lab2(3)]);

% compute the mean values for use later
meanC = (c2+c1)/2;
meanL = (lab2(1)+lab1(1))/2;
[a1,b1] = pol2car([1,h1]);
[a2,b2] = pol2car([1,h2]);
a = (a1+a2)/2;
b = (b1+b2)/2;
[c,meanH] = car2pol([a b]);

% compute the basic delta values
dh = (h2-h1);
if (dh>180)
   dh = dh - 360;
end

dl = lab2(1)-lab1(1);
dc = c2-c1;
dh = 2*sqrt(c1*c2)*sin((dh/2)*pi/180);

T = 1 - 0.17*cos((meanH-30)*pi/180) + 0.24*cos((2*-...
meanH)*pi/180);
T = T + 0.32*cos((3*meanH + 6)*pi/180) - ...
0.20*cos((4*meanH - 63)*pi/180);

dthe = 30*exp(-((meanH-275)/25)^2);
rc = 2*sqrt((meanC^7)/(meanC^7 + 25^7));
rt = -sin(2*dthe*pi/180)*rc;
```

```
Lweight = 1 + (0.015*(meanL-50)^2)/sqrt(20 + (meanL- ...
50)^2);
Cweight = 1 + 0.045*meanC;
Hweight = 1 + 0.015*meanC*T;

dl = dl/(Lweight*paral);
dc = dc/(Cweight*parac);
dh = dh/(Hweight*parah);

de = sqrt(dl^2 + dc^2 + dh^2 + rt*dc*dh);
```

Users may wish to modify the scripts or to convert them into other programming languages. In order to facilitate testing of any implementations of these colour-difference equations Table 5.1 has been provided which lists 10 pairs of samples that Luo *et al.* (2001) have designed for testing the CIEDE2000 equation. The tristimulus values in Table 5.1 are for the 1964 observer and illuminant D65 ($X_n = 94.811$, $Y_n = 100.000$, $Z_n = 107.304$). Table 5.2 lists the colour-difference values for the CIELAB, CMC(1,1), CIE94 and CIEDE2000 equations. Table 5.3 provides more detailed information on the intermediate stages for the CIEDE2000 equation.

Table 5.1 Data for testing implementations of colour-difference equations reproduced from Luo (2001). CIE tristmulus values (illuminant D65 and 1964 observer) are provided for 10 pairs of samples. The standard and trial data are denoted by subscripts S and T, respectively

Pair	X_S	Y_S	Z_S	X_T	Y_T	Z_T
1	19.410000	28.410000	11.576600	19.552500	28.640000	10.579100
2	22.480000	31.600000	38.480000	22.583300	31.370000	36.790100
3	28.995000	29.580000	35.750000	28.770400	29.740000	35.604500
4	4.140000	8.540000	8.030000	4.412900	8.510000	8.645300
5	4.960000	3.720000	19.590000	4.665100	3.810000	17.784800
6	15.600000	9.250000	5.020000	15.914800	9.150000	4.387200
7	73.000000	78.050000	81.800000	73.935100	78.820000	84.515600
8	73.995000	78.320000	85.306000	69.176200	73.400000	79.713000
9	0.704000	0.750000	0.972000	0.613873	0.650000	0.851025
10	0.220000	0.230000	0.325000	0.093262	0.100000	0.145292

Table 5.2 Colour-difference values for the pairs from Table 5.1 computed for the CIELAB, CMC(1:1), CIE94 and CIEDE2000 equations. Note that all the parametric values in CIEDE2000 were set to unity

Pair	CIELAB	CMC(1:1)	CIE94	CIEDE2000
1	3.1819	1.4282	1.3910	1.2644
2	2.2134	1.2549	1.2481	1.2630
3	1.5390	1.7684	1.2980	1.8731
4	4.6063	2.0258	1.8204	1.8645
5	6.5847	3.0870	2.5561	2.0373
6	3.8864	1.7490	1.4249	1.4146
7	1.5051	1.9009	1.4194	1.4440
8	2.3238	1.7026	2.3226	1.5381
9	0.9441	1.8032	0.9385	0.6378
10	1.3191	2.4493	1.3065	0.9082

Table 5.3 Colour-difference values and intermediate values for the pairs from Table 5.1 computed for the CIEDE2000 equation. Note that all the parametric values in CIEDE2000 were set to unity

Pair	G	T	S_L	S_C	S_H	R_T	DE_{00}
1	0.0017	1.3010	1.1427	3.2946	1.9951	0.0000	1.2644
2	0.0490	0.9402	1.1831	2.4549	1.4560	0.0000	1.2630
3	0.4966	0.6952	1.1586	1.3092	1.0717	−0.0032	1.8731
4	0.0063	1.0168	1.2148	2.9105	1.6476	0.0000	1.8645
5	0.0026	0.3636	1.4014	3.1597	1.2617	−1.2537	2.0373
6	0.0013	0.9239	1.1943	3.3888	1.7357	0.0000	1.4146
7	0.4999	1.1546	1.6110	1.1329	1.0511	0.0000	1.4440
8	0.5000	1.3916	1.5930	1.0620	1.0288	0.0000	1.5381
9	0.4999	0.9556	1.6517	1.1057	1.0337	−0.0004	0.6378
10	0.5000	0.7827	1.7246	1.0383	1.0100	0.0000	0.9082

6

Chromatic-adaptation Transforms and Colour Appearance

6.1   Introduction

The distinction between colour specification and colour appearance was touched upon in the review of the CIE system presented in Chapter 1, Section 1.3. Whereas the CIE system of colorimetry (based upon XYZ tristimulus values) is clearly a system for colour specification, some advanced colour specification models such as CIELAB could arguably be described as models of colour appearance. The polar coordinates of CIELAB allow the description of a colour stimulus in terms of three terms, lightness, chroma and hue, and these correlate quite well with the perceptual attributes of brightness, colourfulness and hue. Furthermore, the normalization procedures inherent in the transform from XYZ to $L^*a^*b^*$ result in a^* and b^* values close to zero for a perfect reflecting diffuser (or any surface whose spectral reflectance does not vary with wavelength) irrespective of the illuminant. This is consistent with the fact that surfaces in general tend to retain their colour appearance when viewed under a wide range of light sources and contrasts with the properties of the XYZ system. CIELAB is a relatively poor colour-appearance model, however, and this chapter describes several advanced colour-appearance models (CAMs). The human visual system has a remarkable ability to maintain the colour appearance of an object despite quite large changes in the quality and intensity of the illumination. A white piece of paper tends to look white whether it is viewed by daylight, Tungsten light or candle light. It is generally considered that the human visual system achieves colour constancy by some process that allows it to discount the effect of the illumination. The term chromatic adaptation is often used to describe this process and the chromatic adaptation is said to be complete if the effect of the illumination is completely discounted. Most CAMs therefore include a chromatic-adaptation transform (CAT). A CAT is a method for computing

Computational Colour Science Using MATLAB. By Stephen Westland and Caterina Ripamonti.
© 2004 John Wiley & Sons, Ltd: ISBN 0 470 84562 7

the corresponding colour under a reference illuminant for a stimulus defined under a test illuminant. Corresponding colours are colours that have the same appearance under different illumination.

Fairchild (1998) defines a CAM as any model that includes predictors of at least the relative colour-appearance attributes of lightness, chroma and hue. The attribute brightness is a visual perception according to which an area appears to exhibit more or less light. Lightness is the brightness of an area judged relative to the brightness of a similarly illuminated reference white. The lightness of a sample is in the range 0–100 and is influenced by the surrounding background. Colourfulness is that attribute of a visual sensation according to which an area appears to exhibit more or less chromatic content. Hunt (1952) has shown that the colourfulness of an object increases as the luminance increases so that a typical outdoor scene appears much more colourful in bright sunlight than it does on an overcast day (the Hunt effect). Chroma is the colourfulness of an area judged as a proportion of the brightness of a similarly illuminated reference white. The colourfulness of an area judged in proportion to its brightness is called the saturation. Finally, hue is the attribute of a sensation according to which an area appears to be similar to one, or to a proportion of two, of the perceived colours red, yellow, green and blue.

In this chapter the basic principles that underlie CATs will be introduced and three models (CIECAT94, CMCCAT97 and CMCCAT2000) will be described in detail. The CIECAM97s CAM will then be described and CMCCAM2000 will be introduced. Finally, MATLAB code will be presented for the CATs and CAMs described in this chapter.

6.2 CATs

In psychophysical studies of chromatic adaptation it is useful to define the concept of corresponding colours (colours that have the same appearance under different illumination). In a typical colour-appearance experiment to determine the corresponding colour of a grey surface or chip under a test light source (for example, corresponding to illuminant A) observers adapt to the chip viewed under illuminant A and are then asked to memorize the colour of the chip. The observers are then adapted to the reference light source (often corresponding to illuminant D65) and requested to select a chip, from a large number of different coloured chips, that matches the memorized colour of the original chip that was viewed under the test illumination. If the chip is a perfectly neutral grey, then it would have the chromaticity under D65 corresponding to the D65 illuminant itself and the chromaticity under A corresponding to illuminant A. In Figure 6.1, these points are denoted by an asterix (*) and a cross (+), respectively. If the observer is able to discount the change in illumination perfectly, then the colour appearance of a given surface will be the same under both the test and reference

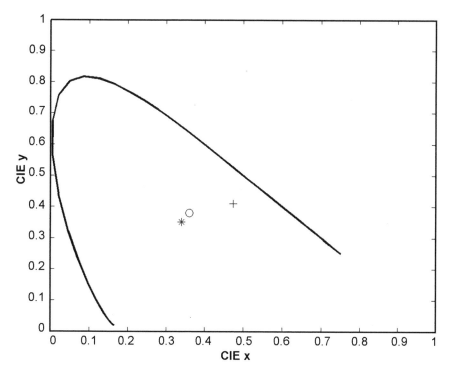

Figure 6.1 Schematic diagram to show the chromaticities of a test stimulus (+) under illuminant A, the stimulus under illuminant D65 (✳) and the hypothetical psychophysically measured corresponding colour (○)

illuminations. Therefore the observer would select the same grey chip under the reference illumination as was viewed under the test illumination. Usually, in such experiments the corresponding colours are not those predicted by a process that discounts the illumination change perfectly. In Figure 6.1 the corresponding colour (○) determined under the reference illuminant (D65) to match a neutral chip viewed under illuminant A (+) is shown for a hypothetical experiment.

Fairchild (1998) has classified chromatic-adaptation mechanisms into two groups: sensory and cognitive. Sensory chromatic-adaptation mechanisms refer to those that respond automatically to the stimulus and are thought to relate to control mechanisms in the sensitivities of the long-, medium- and short-wavelength-sensitive cone classes. Cognitive chromatic-adaptation mechanisms refer to higher level cognitive processes that may relate to our understanding of scene content. Research has shown that chromatic-adaptation mechanisms are quite rapid, being 50% complete after 4 s, 90% complete after 70 s, and 99% complete after 110 s (Fairchild and Lennie, 1992; Fairchild and Reniff, 1995).

A CAT is a method for computing the corresponding colour under a reference illuminant for a stimulus defined under a test illuminant. Most modern CATs are at least loosely based on the von Kries model of adaptation. The likely mechanism underlying this process is that under a reddish light, for example, the long-wavelength-sensitive cones in particular will adapt and so become less sensitive. Under a bluish light, however, the sensitivity of the long-wavelength-sensitive cones will increase. In this way, the idea is that the cone responses for a given surface will stay almost the same even when the illumination is changed, and that the visual system will be able to use the cone excitations to provide a constant appearance for a surface when the illumination changes even though the spectral distribution of light entering the eye is changed.

The changes in sensitivity can be modelled for a static visual system by assuming that the cone responses for a surface under one illuminant can be predicted from those under another illuminant by simple scaling factors. Thus, the long-wavelength response for a surface viewed under one light source can be obtained by multiplying the long-wavelength response for the surface viewed under a different light source by a scalar. The scalar values may be different for each cone class but critically do not depend upon the reflectance or chromaticity of the sample. In terms of linear algebra we can state that the cone responses for (a sample viewed under) one illuminant can be related to those for another illuminant by a linear transform. Since the linear transform's system matrix has non-zero entries only along the major diagonal, it is referred to as a diagonal transform. Thus, for example, the cone responses under one illuminant (represented by the 3×1 column matrix $\mathbf{e}_1$) are related to the cone responses under a second illuminant (represented by the 3×1 column matrix $\mathbf{e}_2$) by the diagonal matrix $\mathbf{D}$, thus

$$\mathbf{e}_2 = \mathbf{D}\mathbf{e}_1, \tag{6.1}$$

where the coefficients of the diagonal matrix are given by the ratios of the long-, medium-, and short-wavelength-sensitive cone responses for a white object viewed under each of the two illuminants,

$$\mathbf{D} = \begin{bmatrix} L_2/L_1 & 0 & 0 \\ 0 & M_2/M_1 & 0 \\ 0 & 0 & S_2/S_1 \end{bmatrix}.$$

The von Kries law is sometimes called the coefficient law or the scaling law since it assumes that the effect of an illumination change can be modelled simply by scaling the tristimulus values or cone responses by scalars or coefficients (the diagonal elements of $\mathbf{D}$). When the von Kries adaptation transform is performed using cone space, Terstiege (1972) has termed this a genuine von Kries transformation, whereas in practice it is often carried out in CIE XYZ space or in an RGB space when it is referred to as a *wrong* von Kries transformation.

Analyses of experimental data suggest that although Equation (6.1) cannot perfectly predict the performance of observers in psychophysical experiments, psychophysical data can be modelled by such a system at least to a first-order approximation (Wandell, 1995). However, Finlayson has shown that a diagonal mapping is always possible between two real three-dimensional spaces ($\Re^3$) if the spaces are first subject to a specific linear transformation. He argues that if the tristimulus values or cone responses are first transformed by a linear transform into a suitable *RGB* space, then a diagonal transform can effectively discount the illumination (Finlayson and Süsstrunk, 2000). The first linear transform is sometimes called a sharp transform since it can be shown to convert the cone responses into a set of channels whose spectral sensitivities are sharper than those that have been measured for humans. We can therefore consider a generalized CAT based upon Equation (6.2) where c_1 and c_2 refer to the tristimulus values of the sample under the two illuminants,

$$c_2 = M_{CAT}^{-1} D\ M_{CAT}\ c_1, \tag{6.2}$$

and the diagonal matrix **D** is now composed from the white points of the two illuminants in the *RGB* space. In Equation (6.2) the tristimulus values are first subject to a linear transform (M_{CAT}) which converts them into *RGB* space and then to a diagonal transform (**D**) to apply the illuminant correction, and finally a linear transform (M_{CAT}^{-1}) to convert back to tristimulus space. Finlayson has derived the *RGB* or sharp transform as given by $M_{CAT} = M_{SHARP}$,

$$M_{SHARP} = \begin{bmatrix} 1.2694 & -0.0988 & -0.1706 \\ -0.8364 & 1.8006 & 0.0357 \\ 0.0297 & -0.0315 & 1.0018 \end{bmatrix}.$$

The most popular CATs are consistent with Finlayson's idea, and the procedure of subjecting the tristimulus values of a stimulus under one illuminant by a 3×3 linear transform M_{CAT}, followed by a diagonal transform **D**, and finally followed by the inverse linear transform M_{CAT}^{-1} to return to tristimulus space is ubiquitous in CAT research. Often researchers refer to the *RGB* space in which the diagonal transform takes place as cone space, although the term is being used loosely in this sense.

A number of CATs are currently in use and most transform the tristimulus values into an *RGB* space before applying the diagonal transform. The *RGB* space differs slightly between the different transforms; that is, the 3×3 linear transform M_{CAT} is different for each CAT. However, more significant differences between the transforms are found in the way in which the elements of the diagonal transform are computed and in which properties of the observing field are used to compute these elements.

Hunt (1998) classified the observing field into five areas: the colour element, the proximal field, the background, the surround and the adapting field, and these areas are shown schematically in Figure 6.2. The colour element is the central area of the observing field and this typically is a uniform patch of approximately $2°$ of visual angle. The proximal field is the immediate

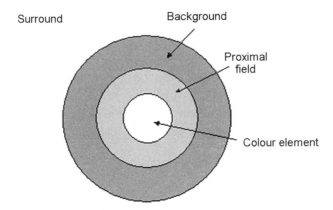

Figure 6.2 Schematic diagram to show the observing field for the description of colour appearance according to Hunt (1998)

environment of the colour element extending for approximately 2° from the edge of the colour element. The colour element and its proximal field are considered to be viewed against the background, a region extending approximately 10° in every direction from the edge of the proximal field. The surround is the field outside of the background. Finally, the adapting field is the total environment within which the colour element, the proximal field and the background are viewed.

It is common practice to follow Moroney's terminology so that the term *adopted white* is used to describe the computational white point used in various model calculations, whereas the term *adapted white* is used to define the white point to which a human observer is considered to be adapted to (Moroney, 2000).

6.2.1 CIECAT94

In 1994 the CIE recommended a CAT developed by Nayatani and his workers (Nayatani *et al.*, 1990, 1999) known as CIECAT94. Unlike the simple von Kries model, CIECAT94 takes into account the luminance level used and the degree of adaptation. This model therefore led the way for a plethora of modern CATs that currently dominate the colour literature. However, a number of studies have shown that the complexity of the CIECAT94 model is not justified by its performance (e.g. Sueeprasan, 2003).

The first stage of the transform is to convert the *XYZ* values of the sample under the test conditions to *RGB* values using a linear transform shown as Equations (6.3):

$$R = 0.40024X + 0.70760Y - 0.08081Z,$$
$$G = -0.22630X + 1.16532Y + 0.04570Z, \qquad (6.3)$$
$$B = 0.00000X + 0.00000Y - 0.91822Z.$$

The corresponding values R_C, G_C and B_C under the reference illuminant are computed according to Equations (6.4),

$$R_C = (Y_0 P_R + n)K^{1/\beta(R_R)}[(R+n)/(Y_0\langle P\rangle + n)]^{\beta(R_T)/\beta(R_R)} - n,$$
$$G_C = (Y_0 Q_R + n)K^{1/\beta(G_R)}[(G+n)/(Y_0\langle Q\rangle + n)]^{\beta(G_T)/\beta(G_R)} - n, \qquad (6.4)$$
$$B_C = (Y_0 S_R + n)K^{1/\beta(B_R)}[(B+n)/(Y_0\langle S\rangle + n)]^{\beta(B_T)/\beta(B_R)} - n,$$

where n is a noise term ($n = 0.1$) and the other parameters are computed according to the following steps:

Step 1: Compute the chromaticity correlates P_T, Q_T, S_T and P_R, Q_R, S_R using

$$P_T = (0.48105x_T + 0.78841y_T - 0.08081)/y_T,$$
$$Q_T = (-0.27200x_T + 1.11962y_T - 0.08081)/y_T$$
$$S_T = (0.48105x_T + 0.78841y_T - 0.08081)/y_T,$$

and

$$P_R = (0.48105x_R + 0.78841y_R - 0.08081)/y_R,$$
$$Q_R = (-0.27200x_R + 1.11962y_R - 0.08081)/y_R,$$
$$S_R = (0.48105x_R + 0.78841y_R - 0.08081)/y_R,$$

where x_R, y_R and x_T, y_T are the chromaticity coordinates of the reference and test illuminants, respectively.

Step 2: Compute the coefficient α for adaptation using

$$\alpha = 0.115\log(L_T) + 0.0025(L^* - 50) + 0.22D + 0.51,$$

where the factor $D = 1.0$ for object colours and $D = 0.0$ for luminous colours (intermediate values of D may be used for projected colour slides) and the value of α is capped to have a maximum $\alpha_{max} = 1.0$. The value of L_T is the luminance (cd/m²) of the adapting test field and L^* is the CIE lightness of the sample under the test illuminant.

Step 3: Compute the adapting chromaticity correlates $\langle P\rangle$, $\langle Q\rangle$, and $\langle S\rangle$ using

$$\langle P\rangle = \alpha P_T + (1-\alpha)P_R,$$
$$\langle Q\rangle = \alpha Q_T + (1-\alpha)Q_R,$$
$$\langle S\rangle = \alpha S_T + (1-\alpha)S_R.$$

Step 4: Compute the effective adapting responses of the test (R_T, G_T, B_T) and reference (R_R, G_R, B_R) conditions using

$$R_T = L_T \langle P \rangle,$$
$$G_T = L_T \langle Q \rangle,$$
$$B_T = L_T \langle S \rangle,$$

and

$$R_R = L_R P_R,$$
$$G_R = L_R Q_R,$$
$$B_R = L_R S_R.$$

Step 5: Compute the exponents of the red, green and blue transformations $\beta(R_T)$, $\beta(G_T)$, $\beta(B_T)$ and $\beta(R_R)$, $\beta(G_R)$, $\beta(B_R)$ using

$$\beta(R_T) = (6.469 + 6.326 R_T^{0.4495})/(6.469 + R_T^{0.4495}),$$
$$\beta(G_T) = (6.469 + 6.326 G_T^{0.4495})/(6.469 + G_T^{0.4495}),$$
$$\beta(R_R) = (6.469 + 6.326 R_R^{0.4495})/(6.469 + R_R^{0.4495}),$$
$$\beta(G_R) = (6.469 + 6.326 G_R^{0.4495})/(6.469 + G_R^{0.4495}),$$
$$\beta(B_T) = 0.7844(8.414 + 8.091 B_T^{0.5128})/(8.414 + B_T^{0.5128}),$$
$$\beta(B_R) = 0.7844(8.414 + 8.091 B_R^{0.5128})/(8.414 + B_R^{0.5128}).$$

Step 6: Compute the coefficient K using

$$K = p/q,$$

where

$$p = [(Y \langle P \rangle + n)/(20 \langle P \rangle + n)]^{2\beta(R_T)/3}[(Y \langle Q \rangle + n)/(20 \langle Q \rangle + n)]^{\beta(G_T)/3}$$

and

$$q = [(Y P_R + n)/(20 P_R + n)]^{2\beta(R_R)/3}[(Y Q_R + n)/(20 Q_R + n)]^{\beta(G_R)/3},$$

and where Y is the luminance factor (per cent) of the test adapting field.

Finally, the inverse of Equations (6.3) is used to transform the correlated RGB values into correlated XYZ values,

$$X_R = 1.85995 R_R - 1.12939 G_R + 0.21990 B_R,$$
$$Y_R = 0.36119 R_R + 0.63881 G_R + 0.00000 B_R, \qquad (6.5)$$
$$Z_R = 0.00000 R_R + 0.00000 G_R + 1.08906 B_R.$$

6.2.2 CMCCAT97

The CMCCAT97 CAT (Luo and Hunt, 1998b) is a modified version of the Bradford (BFD) CAT (Lam, 1985) and is used in the CIECAM97s CAM. It is essentially based on the form of Equation (6.2) where $\mathbf{M}_{CAT} = \mathbf{M}_{BFD}$, and

$$\mathbf{M}_{BFD} = \begin{bmatrix} 0.8951 & 0.2664 & -0.1614 \\ -0.7502 & 1.7135 & 0.0367 \\ 0.0389 & -0.0685 & 1.0296 \end{bmatrix}.$$

However, in CMCCAT97 the transformation to RGB space is performed upon the tristimulus values normalized by the Y value of the sample. Thus,

$$\begin{aligned} R &= 0.8951X/Y + 0.2664Y/Y - 0.1614Z/Y, \\ G &= -0.7502X/Y + 1.7135Y/Y + 0.0367Z/Y, \\ B &= 0.0389X/Y - 0.0685Y/Y + 1.0296Z/Y, \end{aligned} \tag{6.6}$$

where X, Y and Z are the tristimulus values of the sample under the test illuminant.

Note that this stage is equivalent to multiplying the 3×1 column matrix of normalized tristimulus values by the 3×3 matrix $\mathbf{M}_{BFD}$. The XYZ values of the adopted test and reference illuminants must also be subject to the generic form of Equations (6.6) to yield R_{WT}, G_{WT}, B_{WT}, and R_{WR}, G_{WR}, B_{WR}, respectively. To avoid confusion, the equations for these transforms are given in full by Equations (6.7) and (6.8),

$$\begin{aligned} R_{WT} &= 0.8951X_{WT}/Y_{WT} + 0.2664Y_{WT}/Y_{WT} - 0.1614Z_{WT}/Y_{WT}, \\ G_{WT} &= -0.7502X_{WT}/Y_{WT} + 1.7135Y_{WT}/Y_{WT} + 0.0367Z_{WT}/Y_{WT}, \\ B_{WT} &= 0.0389X_{WT}/Y_{WT} - 0.0685Y_{WT}/Y_{WT} + 1.0296Z_{WT}/Y_{WT}, \end{aligned} \tag{6.7}$$

and

$$\begin{aligned} R_{WR} &= 0.8951X_{WR}/Y_{WR} + 0.2664Y_{WR}/Y_{WR} - 0.1614Z_{WR}/Y_{WR}, \\ G_{WR} &= -0.7502X_{WR}/Y_{WR} + 1.7135Y_{WR}/Y_{WR} + 0.0367Z_{WR}/Y_{WR}, \\ B_{WR} &= 0.0389X_{WR}/Y_{WR} - 0.0685Y_{WR}/Y_{WR} + 1.0296Z_{WR}/Y_{WR}, \end{aligned} \tag{6.8}$$

where X_{WT}, Y_{WT}, Z_{WT} and X_{WR}, Y_{WR}, Z_{WR} are the tristimulus values of the adopted test and reference illuminants, respectively.

CMCCAT97 incorporates the degree of adaptation D and this is computed using Equation (6.9),

$$D = F - F/[1 + 2(L_A^{1/4}) + L_A^2/300], \tag{6.9}$$

where $F = 1$ for surfaces seen under typical viewing conditions, $F = 0.9$ for surfaces seen under dim or dark conditions and L_A is the luminance (cd/m²) of the adapting test field. The degree of adaptation D is then used with the ratios of the white points of the illuminants to convert the *RGB* values of the sample to the *RGB* values of the corresponding colour,

$$R_C = [D(R_{WR}/R_{WT}) + 1 - D]R,$$
$$G_C = [D(G_{WR}/G_{WT}) + 1 - D]G, \tag{6.10}$$

$$B_C = \begin{cases} [D(B_{WR}/B_{WT}^P) + 1 - D]B^P, & \text{if } B_C > 0, \\ -[D(B_{WR}/B_{WT}^P) + 1 - D]|B|^P, & \text{otherwise,} \end{cases}$$

where

$$P = (B_{WT}/B_{WR})^{0.0834}.$$

Note that when $D = 1$, the transform [Equations (6.10)] is quite close to the form of a von Kries or diagonal transform except that a non-linearity is applied to the *B* channel.

Finally, the corresponding *RGB* values are converted back to tristimulus values by multiplying them by the inverse of $\mathbf{M}_{BFD}$ to yield the normalized tristimulus values which finally can be converted by multiplying each by the *Y* tristimulus value of the sample under the test illuminant,

$$X_C = Y(0.9870R_C - 0.1471G_C + 0.1600B_C),$$
$$Y_C = Y(0.4323R_C + 0.5184G_C + 0.0493B_C), \tag{6.11}$$

and

$$Z_C = Y(-0.0085R_C + 0.0400G_C + 0.9685B_C).$$

6.2.3 CMCCAT2000

There is some uncertainty over the reversibility of the CMCCAT97 transform which arises because of the power function in Equations (6.10). Although this problem has been solved by a small revision (Li *et al.*, 2000) a further weakness of the CMCCAT97 is that it was derived by fitting only a relatively small data set. Further work resulted in the development of a new CAT that was accepted by the Colour Measurement Committee and known as CMCCAT2000 (Li *et al.*, 2002). In CMCCAT2000 the power function was removed so that the transform is fully reversible and the model was fitted to all available data sets. Consequently the linear transform component, $\mathbf{M}_{CMCCAT2000}$, of CMCCAT2000 is slightly different from that of CMCCAT97 and is shown below,

$$\mathbf{M}_{CMCCAT2000} = \begin{bmatrix} 0.7982 & 0.3389 & -0.1371 \\ -0.5918 & 1.5512 & 0.0406 \\ 0.0008 & 0.0239 & 0.9753 \end{bmatrix}.$$

The first step in CMCCAT2000 is to transform the tristimulus values of the sample under the test illuminant to RGB values using Equations (6.12),

$$R = 0.7982X/Y + 0.3389Y/Y - 0.1371Z/Y,$$
$$G = -0.5918X/Y + 1.5512Y/Y + 0.0406Z/Y,$$

(6.12)

and

$$B = 0.0008X/Y + 0.0239Y/Y + 0.9753Z/Y,$$

where X, Y and Z are the tristimulus values of the sample. Again, note that the normalized tristimulus values are used as the input to the linear transform [Equations (6.12)]. Similar transforms are computed for the test and reference illuminants to produce R_{WT}, G_{WT}, B_{WT} and R_{WR}, G_{WR}, B_{WR}, respectively.

Note that this stage is equivalent to multiplying the 3×1 column matrix of normalized tristimulus values by the 3×3 matrix $\mathbf{M}_{CMCCAT2000}$. CMCCAT2000 incorporates the degree of adaptation D and this is computed by Equation (6.13),

$$D = F\{0.08 \log[0.5(L_{AT} + L_{AR})] + 0.76 - 0.45(L_{AT} - L_{AR})/(L_{AT} + L_{AR})\},$$

(6.13)

where the parameter $F = 1.0$ for average viewing conditions, $F = 0.8$ for dim and dark surround conditions and L_{AT} and L_{AR} are the luminances of the test and reference adapting fields, respectively (note that CMCCAT97 did not account for the luminance of the reference field).

The degree of adaptation D is then used to convert the RGB values of the sample to the RGB values of the corresponding colour,

$$R_C = [D(R_{WR}/R_{WT}) + 1 - D]R,$$
$$G_C = [D(G_{WR}/G_{WT}) + 1 - D]G,$$
$$B_C = [D(B_{WR}/B_{WT}) + 1 - D]B.$$

(6.14)

Note that when $D = 1$, the transform [Equations (6.14)] is simply a diagonal transform of the RGB values.

Finally, the corresponding RGB values are converted back to tristimulus values by multiplying them by the inverse of $\mathbf{M}_{CMCCAT2000}$. This procedure yields the normalized tristimulus values which finally can be converted by multiplying each by the Y tristimulus value of the sample under the test illuminant,

$$X_C = Y(1.0765R_C - 0.2377G_C + 0.1612B_C),$$
$$Y_C = Y(0.4110R_C + 0.5543G_C + 0.0347B_C),$$
$$Z_C = Y(-0.0110R_C - 0.0134G_C + 1.0243B_C).$$

(6.15)

A slight revision of CMCCAT2000, known as CAT02, has been developed whereby the matrix $\mathbf{M}_{CAT02}$ was derived by fitting all the available data sets apart from one (which was excluded on the basis that it refers to narrowband light sources that are unlikely to be encountered in practical situations).

6.3 CAMs

According to Fairchild's definition of a CAM, CIELAB can be considered to be a CAM but it makes relatively poor predictions of colour appearance in most cases. The main problem of using CIELAB as a CAM is demonstrated by the simple example where a grey patch is shown surrounded by either a white background or a black background (Figure 6.3).

The CIELAB colour coordinates for the two grey patches shown in Figure 6.3 are identical but the colour appearances of the two grey patches are quite different. Although the normalization of the tristimulus values by the white point in the computation of CIELAB values attempts to deal with some issues of colour appearance (those caused by the ability of the visual system to compensate for changes in the illumination) it does not contain any spatial component. Yet in everyday viewing the colours that we see are almost always related colours. That is, we see colours in relation to the surrounding colours in a scene. A CAM should, for example, be able to predict an increase in lightness when a grey paper is viewed against a dark background compared with when it is viewed against a light background. For many technologies, of course, colour appearance is not important and basic colorimetry is sufficient (Berns, 2000). So, for example, if we wish to compare a trial fabric sample with a standard we are often only concerned with whether the trial matches the standard rather than with what the two samples actually look like. The importance of this distinction (see Westland, 2002) is critical if the recent work on colour appearance is to be understood.

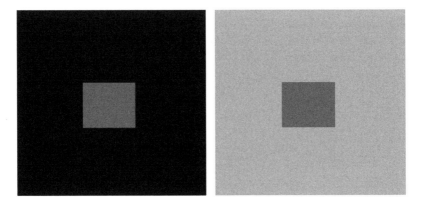

Figure 6.3 The grey patch looks lighter on the dark background than it does on the light background

6.3.1 CIECAM97s

In 1997 the CIE Technical Committee TC1-34 agreed to adopt a CAM known as CIECAM97s (Luo and Hunt, 1998a). CIECAM97s comprises two parts: the first part is a CAT that computes the corresponding tristimulus values; the second part calculates a set of colour-appearance descriptors for the corresponding tristimulus values. The particular CAT that is used in CIECAM97s is known as CMCCAT97 (Luo and Hunt, 1998b) and this is described in Section 6.2.2. CIECAM97s includes a forward and a reverse mode. The forward mode transforms the tristimulus values of a sample under a non-daylight illuminant, such as A, to those for the corresponding colour under a daylight illuminant (in fact, the equal-energy illuminant) and then computes some terms that describe the colour appearance of the sample under the daylight illuminant. The reverse mode is used to predict the tristimulus values under a non-daylight illuminant based upon the colour-appearance descriptors for a sample viewed in daylight.

The output of the forward mode is a set of attributes that predict colour appearance: brightness, lightness, colourfulness, chroma, saturation and hue.

The predictions from the CIECAM97s model are in agreement with a number of colour-appearance phenomena such as chromatic adaptation (CMCCAT97 is included within CIECAM97s), Hunt's effect, Stevens' effect (Stevens and Stevens, 1963) and the surround effect (Bartleson and Breneman, 1967). However, there are certain well-known colour-appearance phenomena that CIECAM97s cannot predict (Luo, 2002b).

The forward mode considers a sample viewed under a (non-white) test illuminant against an achromatic background and computes the corresponding colour-appearance attributes for the sample under a (white) reference illuminant. The starting data required for the forward transform include the tristimulus values of the sample in the test illuminant (X, Y, Z), the adopted white in the test illuminant (X_{WT}, Y_{WT}, Z_{WT}), the reference white in the reference illuminant (X_{WR}, Y_{WR}, Z_{WR}), the Y tristimulus value of the achromatic background against which the patch is viewed (Y_B), and the luminance (in units of cd/m²) of the reference white (L_W) and the achromatic background (L_A) against which the sample is viewed.

The model requires that values are assigned to four parameters, F, c, F_{LL} and N_C. Different values are recommended for these parameters depending upon the viewing conditions (Luo, 2002b). However, for large moderately illuminated scenes the following values should be used: $F = 1.0$, $c = 0.690$, $F_{LL} = 0.0$ and $N_C = 1.0$. The following steps describe in detail the computation of the forward mode of CIECAM97s.

Step 1: Calculate the *RGB* values for the test sample (R, G, B) and for the reference white under the test (R_{WT}, G_{WT}, B_{WT}) and reference fields (R_{WR}, G_{WR}, B_{WR}) using Equations (6.6). Note that this stage is equivalent to multiplying the

3×1 column matrix of normalized tristimulus values by the 3×3 matrix $\mathbf{M}_{BFD}$. The XYZ values of the sample are used with Equations (6.6) to compute the RGB values for the sample; X_{WT}, Y_{WT}, Z_{WT} and X_{WR}, Y_{WR}, Z_{WR} are used to compute R_{WT}, G_{WT}, B_{WT} and R_{WR}, G_{WR}, B_{WR}, respectively.

Step 2: Calculate the corresponding RGB values for the test sample (R_C, G_C, B_C) and for the reference white (R_{WC}, G_{WC}, B_{WC}) using Equations (6.10).

Step 3: Calculate the luminance level adaptation factor (F_L), the chromatic background induction factor (N_{CB}) and the brightness background induction factor (N_{BB}),

$$F_L = K^4(L_A) + 0.1(1 - K^4)^2(5L_A)^{1/3}, \tag{6.16}$$

where $K = 1/(5L_A + 1)$, $N_{CB} = N_{BB} = 0.725(1/n)^{0.2}$ and $n = Y_B/Y_W$.

Step 4: Calculate the corresponding tristimulus values for the test sample (R', G', B') and for the reference white (R'_W, G'_W, B'_W),

$$\begin{bmatrix} R' \\ G' \\ B' \end{bmatrix} = \mathbf{M}_H \mathbf{M}_{BFD}^{-1} \begin{bmatrix} R_C Y \\ G_C Y \\ B_C Y \end{bmatrix}, \tag{6.17}$$

where

$$\mathbf{M}_{BFD}^{-1} = \begin{bmatrix} 0.9870 & -0.1471 & 0.1600 \\ 0.4323 & 0.5184 & 0.0493 \\ -0.0085 & 0.0400 & 0.9685 \end{bmatrix}$$

and

$$\mathbf{M}_H = \begin{bmatrix} 0.38971 & 0.68898 & -0.07868 \\ -0.22981 & 1.18340 & 0.04641 \\ 0.00000 & 0.00000 & 1.00000 \end{bmatrix}.$$

Step 5: Calculate the cone responses after adaptation for the test sample (R'_a, G'_a, B'_a) and for the reference white (R'_{aW}, G'_{aW}, B'_{aW}),

$$R'_a = 1 + [40(F_L R'/100)^{0.73}]/[(F_L R'/100)^{0.73} + 2],$$
$$G'_a = 1 + [40(F_L G'/100)^{0.73}]/[(F_L G'/100)^{0.73} + 2], \tag{6.18}$$
$$B'_a = 1 + [40(F_L B'/100)^{0.73}]/[(F_L B'/100)^{0.73} + 2],$$

and where, if $R'_a < 0$,

$$R'_a = 1 - [40(-R'/100)^{0.73}]/[(-R'/100)^{0.73} + 2].$$

Table 6.1 Values of H, h and e for the unique hues

	Red	Yellow	Green	Blue	Red
H	0	100	200	300	400
h	20.14	90.00	164.25	237.53	380.14
e	0.8	0.7	1.0	1.2	0.8

Step 6: Calculate the red-green (a) and yellow-blue (b) opponent correlates,

$$a = R'_a - 12G'_a/11 + B'_a/11,$$
$$b = (R'_a + G'_a - 2B'_a)/9. \tag{6.19}$$

Step 7: Calculate the hue angle (h),

$$h = \tan^{-1}(b/a)\,(180/\pi). \tag{6.20}$$

Step 8: Calculate the eccentricity factor (e) and the hue quadrature (H),

$$H = H_1 + [100(h - h_1)/e_1]/[(h - h_1)/e_1 + (h_2 - h_1)/e_2],$$
$$e = e_1 + (e_2 - e_1)(h - h_1)/(h_2 - h_1), \tag{6.21}$$

where H_1 is either 0, 100, 200 or 300 depending upon whether red, yellow, green or blue respectively, is the hue having the nearest lower value of h. Table 6.1 shows the values of H, h and e for the unique hues.

The values of e_1 and h_1 are the values of e and h for the unique hue having the nearest lower value of h; the values of e_2 and h_2 are the values of e and h for the unique hue having the nearest higher value of h.

Step 9: Calculate the achromatic response of the sample (A) and of the reference white (A_W),

$$A = [2R'_a + G'_a + B'_a/20 - 2.05]N_{BB},$$
$$A_W = [2R'_{aW} + G'_{aW} + B'_{aW}/20 - 2.05]N_{BB}. \tag{6.22}$$

Step 10: Calculate the lightness of the sample (J),

$$J = 100(A/A_W)^{cz}, \tag{6.23}$$

where $z = 1 + F_{LL}n^{1/2}$.

Step 11: Calculate the brightness of the sample (Q),

$$Q = (1.24/c)(J/100)^{0.67}(A_W + 3)^{0.9}. \tag{6.24}$$

Step 12: Calculate the saturation of the sample (s),

$$s = [5000(a^2 + b^2)^{1/2}10eN_CN_{BB}/13]/[R'_a + G'_{aW} + 21B'_a/20]. \tag{6.25}$$

Step 13: Calculate the chroma of the sample (C),

$$C = 2.44s^{0.69}(J/100)^{0.67n}(1.64 - 0.29^n). \tag{6.26}$$

Step 14: Calculate the colourfulness of the sample (M),

$$M = CF_L^{0.15}. \tag{6.27}$$

6.3.2 CMCCAM2000

Although CIECAM97s is widely used in the colour-management industry a number of alternative models have been produced. Currently there is much focus on the nature of the CAT that should be used (recall that CIECAM97s uses CMCCAT97). Nayatani *et al.* (1999) have recently proposed an alternative CAT and a further transform $\mathbf{M}_{SHARP}$ has been developed based directly upon the principle of chromatic sharpening (Finlayson and Süsstrunk, 2000). Luo and his colleagues (Li *et al.*, 2002) have developed CMCCAT2000 and this has been adopted by the Colour Measurement Committee (CMC) of the Society of Dyers and Colourists. It has been claimed that CMCCAT2000 gives a prediction to almost all of the available data sets that is more accurate than any of the other published transforms (Li *et al.*, 2002). CMCCAT2000 is the CAT that forms the basis of a colour-appearance model known as CMCCAM2000.

6.4 Implementations and examples

6.4.1 CATs

The function *cmccat97* implements CMCCAT97 and the format for the function is

```
[xyzc] = cmccat97(xyz, xyzt, xyzr, la, f)
```

where **xyz** is a 3×1 matrix of the tristimulus values for the sample under the test illuminant, **xyzt** and **xyzr** are 3×1 matrices whose entries hold the white points of the adopted test and reference illuminants, respectively, and **la** and **f** are parameters (both 1×1). The parameter **la** holds the luminance of the adapting test field and this has a default value of 100 cd/m^2. The parameter **f** has a default value of 1.0 and this corresponds to typical viewing conditions (a value of 0.9 should be used for dark or dim conditions).

Box 15: *cmccat97.m*

```
function [xyzc] = cmccat97(xyz,xyzt,xyzr,la,f)

% function [xyzc] = cmccat97(xyz,xyzt,xyzr,la,f)
% implements the CMCCAT97 chromatic adaptation transform
% operates on 1 by 3 matrix xyz containing tristimulus
% values of the stimulus under the test illuminant
% xyzt and xyzr are 1 by 3 matrices containing the
% white points for the test and reference conditions
% f has default value 1
% la is the luminance of the adapting test field
% and has default value of 100
% xyzc contains the tristimulus values of the
% stimulus under the reference illuminant
% see also cmccat00

% check the arguments
xyz = xyz(:); % force to be a column matrix
xyzt = xyzt(:); % force to be a column matrix
xyzr = xyzr(:); % force to be a column matrix

if (length(xyz) ~= 3)
  disp('first argument must be 3 by 1 or 1 by 3'); return;
end
if (length(xyzt) ~= 3)
  disp('second argument must be 3 by 1 or 1 by 3'); return;
end
if (length(xyzr) ~= 3)
  disp('third argument must be 3 by 1 or 1 by 3'); return;
end

if (nargin<4)
  disp('using default values of LA and F')
  la = 100.0; f = 1.0;
end

% define the matrix for the transform to 'cone' space
M(1,:) = [0.8951 0.2664 -0.1614];
M(2,:) = [-0.7502 1.7135 0.0367];
M(3,:) = [0.0389 -0.0685 1.0296];
```

```
% implement step 1: normalise xyz and transform to rgb
rgb = M*(xyz/xyz(2));
rgbt = M*(xyzt/xyzt(2));
rgbr = M*(xyzr/xyzr(2));

% implement step 2: compute d, the degree of adaptation
d = f - f/(1 + 2*(la^0.25) + la*la/300);
% clip d if it is outside the range [0,1]
if (d<0)
  d = 0;
elseif (d>1)
  d = 1;
end

p = (rgbt(3)/rgbr(3))^0.0834;

% implement step 3: compute corresponding rgb values
rgbc(1) = rgb(1)*(d*rgbr(1)/rgbt(1) + 1 - d);
rgbc(2) = rgb(2)*(d*rgbr(2)/rgbt(2) + 1 - d);
rgbc(3) = (d*(rgbr(3)/(rgbt(3)^p)) + 1 -
d)*abs(rgb(3))^p;
if (rgb(3) < 0)
  rgbc(3) = -rgbc(3);
end

% implement step 4: convert from rgb to xyz
xyzc = inv(M)*(rgbc'*xyz(2));
```

The default values of the parameters will be used if the following form of the function call is used:

```
[xyzc] = cmccat97(xyz, xyzt, xyzr).
```

The function returns a 3×1 matrix **xyzc** containing the corresponding tristimulus value for the sample under the reference illuminant. The corresponding colour under illuminant D65 for a sample with tristimulus values of 34.1827, 39.2556 and 14.8082 under illuminant A would be generated using the following MATLAB code,

```
xyz = [34.1827 39.2556 14.8082];
xyzt = [111.144 100.00 35.200];
```

```
xyzr = [94.811 100.00 107.304];
[xyzc] = cmccat97(xyz, xyzt, xyzr, 100.0, 1.0).
```

The corresponding tristimulus values that result from this computation are displayed along with test and corresponding values for three other samples in Table 6.2 which is provided to allow programmers to quickly check the fidelity of their own implementations of CMCCAT97.

Table 6.2 Data for testing implementations of CMCCAT97. Tristimulus values (XYZ) under illuminant A are shown for four Munsell surfaces with the corresponding values $(X_C Y_C Z_C)$ for illuminant D65 for the 1964 standard observer. The parameters L_A and F are fixed at 100 and 1, respectively

X	34.1827	52.1707	11.9082	43.4214
Y	39.2556	35.6248	11.3118	43.2444
Z	14.8082	2.0716	12.0749	4.1055
D	0.9754	0.9754	0.9754	0.9754
p	0.9080	0.9080	0.9080	0.9080
X_C	29.7637	40.6782	12.7074	33.2293
Y_C	40.7096	32.5784	12.2382	42.5732
Z_C	43.6411	6.8944	33.4816	11.9849

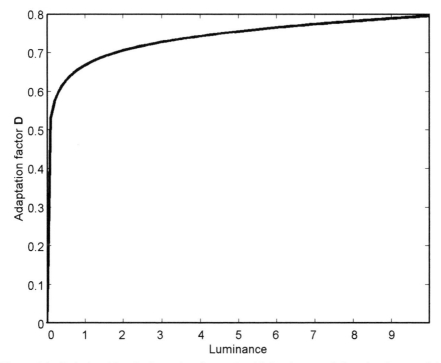

Figure 6.4 Relationship of adaptation factor D with luminance of the adapting test field for CMCCAT97

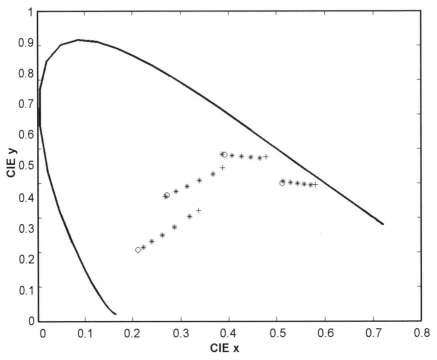

Figure 6.5 Chromaticities for four Munsell samples under illuminant A ($+$) and D65 ($\circ$) with corresponding colours predicted for D65 from A based upon CMCCAT97 for varying degrees of adaption ($*$)

Note that when the luminance L_A of the adapting test field is $100\,\text{cd/m}^2$ the degree of adaptation D is computed as 0.9080. In fact, the adaptation factor is related to L_A as shown in Figure 6.4.

Figure 6.5 illustrates the effect of the degree of adaptation on the corresponding colours predicted by CMCCAT97 for the samples from Table 6.2. The $+$ and $\circ$ symbols represent the chromaticities of the samples for illuminants A and D65, respectively, whereas the $*$ represents the corresponding colours predicted for D65 from the chromaticities under A for various degrees of adaptation.

The function *cmccat00* implements CMCCAT2000 and the format for the function is

```
[xyzc] = cmccat00(xyz, xyzt, xyzr, lt, lw, f),
```

where **xyz** is a 3×1 matrix of the tristimulus values for the sample under the test illuminant, **xyzt** and **xyzr** are the white points of the adopted test and reference illuminants, respectively, and **lt**, **lw** and **f** are parameters. The parameters **lt** and **lw** hold the luminances of the adapting test and reference fields, respectively, and have default values of 100 cd/m². The parameter **f** has a default value of 1.0 and this corresponds to typical viewing conditions (a value of 0.9 should be used for dark or dim conditions).

Box 16: *cmccat00.m*

```
function [xyzc] = cmccat00(xyz,xyzt,xyzr,lt,lw,f)

% function [xyzc] = cmccat00(xyz,xyzt,xyzr,lt,lw,f)
% implements CMCCAT2000 chromatic adaptation transform
% operates on 1 by 3 matrix xyz containing tristimulus
% values of the stimulus under the test illuminant
% xyzt and xyzr are 1 by 3 matrices containing the
% white points for the test and reference conditions
% f has default value 1
% lt is the luminance of the adapting test field
% and has default value of 100
% lw is the luminance of the adapting reference field
% and has default value of 100
% xyzc contains the tristimulus values of the
% stimulus under the reference illuminant

% check the arguments
xyz = xyz(:); % force to be a column matrix
xyzt = xyzt(:); % force to be a column matrix
xyzr = xyzr(:); % force to be a column matrix

if (length(xyz) ~= 3)
   disp('first argument must be 3 by 1 or 1 by 3'); return;
end
if (length(xyzt) ~= 3)
   disp('second argument must be 3 by 1 or 1 by 3'); return;
end
if (length(xyzr) ~= 3)
   disp('third argument must be 3 by 1 or 1 by 3'); return;
end

if (nargin<4)
   disp('using default values of lt, lw and f')
   lt = 100.0; lw = 100.0; f = 1.0;
end
```

```
% define the matrix for the transform to 'cone' space
M(1,:) = [0.7982 0.3389 -0.1371];
M(2,:) = [-0.5918 1.5512 0.0406];
M(3,:) = [0.0008 0.0239 0.9753];

% implement step 1: normalise xyz and transform to rgb
rgb = M*(xyz/xyz(2));
rgbt = M*(xyzt/xyzt(2));
rgbr = M*(xyzr/xyzr(2));

% implement step 2: compute d, the degree of adaptation
d = f*(0.08*log10(lt+lw) + 0.76 - 0.45*(lt-lw)/(lt+lw));
% clip d if it is outside the range [0,1]
if (d<0)
  d = 0;
elseif (d>1)
  d = 1;
end

% implement step 3: compute corresponding rgb values
rgbc(1) = rgb(1)*(d*rgbr(1)/rgbt(1) + 1 - d);
rgbc(2) = rgb(2)*(d*rgbr(2)/rgbt(2) + 1 - d);
rgbc(3) = rgb(3)*(d*rgbr(3)/rgbt(3) + 1 - d);

% implement step 4: convert from rgb to xyz
xyzc = inv(M)*(rgbc'*xyz(2));
```

The default values of the parameters will be used if the following form of the function call is used,

```
[xyzc] = cmccat00(xyz, xyzt, xyzr)
```

The degree of adaptation depends upon the luminances of both the test and reference illuminants as illustrated by Figure 6.6, which was generated by the following MATLAB code:

```
f = 1;
x = linspace(0,10,21); % variable for lt
y = linspace(0,10,21); % variable for lw
z = zeros(21,21);
```

```
for j = 1:21
  for i = 1:21

    lt = x(i);
    lw = y(j);
    d = f*(0.08*log10(lt+lw) + 0.76 - 0.45*(lt-lw)/
    (lt+lw));
    if (d>1)
      d = 1;
    end
    z(i,j) = d;
  end
end

z(1,1) = 1;
surf(x,y,z)
colormap('grey')
```

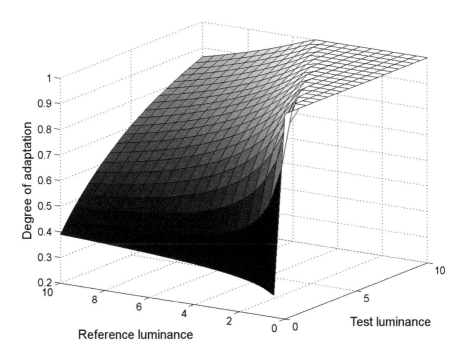

Figure 6.6 The dependence of the degree of adaptation on the adopted test and reference luminance values for CMCCAT2000

Table 6.3 Data for testing implementations of CMCCAT2000. Tristimulus values (XYZ) under illuminant A are shown for four Munsell surfaces with the corresponding values ($X_C Y_C Z_C$) for illuminant D65 for the 1964 standard observer. The parameters L_T, L_W and F are fixed at 100, 100, and 1, respectively

X	34.1827	52.1707	11.9082	43.4214
Y	39.2556	35.6248	11.3118	43.2444
Z	14.8082	2.0716	12.0749	4.1055
D	0.9441	0.9441	0.9441	0.9441
X_C	29.7512	41.6879	12.7603	33.9602
Y_C	40.3041	33.5491	11.9818	42.8973
Z_C	43.2750	7.5182	34.3499	13.5004

The samples that were used to test CMCCAT97 (Table 6.2) have also been used to test CMCCAT2000 (Table 6.3) for comparison.

6.4.2 Computing colour appearance

The forward mode of the CIECAM97s CAM has been implemented by the function *ciecam97s*. The format for this function is

$$[j,c,h,m,hq,s,q,cd] = ciecam97s(xyz,xyzw,la,yb,p),$$

where **xyz** and **xyzw** are 3×1 column matrices containing the tristimulus values of the colour stimulus and the adopted white, respectively, under the test illuminant, **la** is the luminance of the achromatic background in units of cd/m², **yb** is the Y value of achromatic background, and the parameter **p** is a 4×1 column matrix containing the four surround parameters F, c, F_{LL} and N_C. The *ciecam97s* function outputs eight variables. The first seven variables correspond to the lightness (**j**), chroma (**c**), hue quadrature (**hq**), colourfulness (**m**), hue angle (**h**), Saturation (**s**) and brightness (**q**) of the colour stimulus under the reference illuminant. The eighth output parameter, **cd**, is a string that contains the colour descriptor.

Box 17: *ciecam97s.m*

```
function [j,c,hq,m,h,s,q,cd] = ciecam97s(xyz,xyzw,
la,yb,para)

% function [j,c,hq,m,h,s,q,cd] = ciecam97s(xyz,xyzw,
% la,yb,para)
```

```
% implements the CIECAM97s colour appearance model
% operates on 1 by 3 matrix xyz containing tristimulus
% values of the stimulus under the test illuminant
% xyzt and xyzr are 1 by 3 matrices containing the
% white points for the test and reference conditions
% la and yb are the luminance and Y tristimulus values of
% the achromatic background against which the sample is
% viewed
% para is a 1 by 4 matrix containing c, Nc, Fll and F

c = para(1); nc = para(2);
fll = para(3); f = para(4);
MH = [0.38971 0.68898 -0.07868; -0.22981 1.18340 0.04641;
0.0 0.0 1.0];
MBFD = [0.8951 0.2664 -0.1614; -0.7502 1.7135 0.0367;
0.0389 -0.0685 1.0296];
% white in reference light
x = 0.3333; y = 0.3333; Y = 100.00;
xyzwr(1) = x*Y/y; xyzwr(2) = Y; xyzwr(3) = (1-x-y)*Y/y;
xyzwr = xyzwr';

% step 1
rgb = MBFD*(xyz/xyz(2));
rgbw = MBFD*(xyzw/xyzw(2));
rgbwr = MBFD*(xyzwr/xyzwr(2));

% step 2
d = f - f/(1 + 2*(la^0.25) + (la^2)/300);
p = (rgbw(3)/rgbwr(3))^0.0834;

rgbc(1) = (d*(rgbwr(1)/rgbw(1)) + 1 - d)*rgb(1);
rgbc(2) = (d*(rgbwr(2)/rgbw(2)) + 1 - d)*rgb(2);
rgbc(3) = (d*(rgbwr(3)/(rgbw(3)^p)) + 1 - d)*
abs(rgb(3))^p; if (rgb(3) <0)
  rgbc(3) = -rgbc(3);
end

rgbwc(1) = (d*(rgbwr(1)/rgbw(1)) + 1 - d)*rgbw(1);
rgbwc(2) = (d*(rgbwr(2)/rgbw(2)) + 1 - d)*rgbw(2);
rgbwc(3) = (d*(rgbwr(3)/(rgbw(3)^p)) + 1 - d)* ...
abs(rgbw(3))^p; if (rgbw(3) <0)
  rgbwc(3) = -rgbwc(3);
```

```
end
% step 3
k = 1/(5*la+1);
n = yb/xyzw(2);
ncb = 0.725*(1/n)^0.2;
nbb = ncb;
fl = (k^4)*la+0.1*((1-k^4)^2)*((5*la)^(1/3)));

% step 4
rgbp = MH*inv(MBFD)*(rgbc*xyz(2))';
rgbpw = MH*inv(MBFD)*(rgbwc*xyzw(2))';

% step 5
rgbpa = zeros(3,1);
for i = 1:3
  x = fl*rgbp(i)/100;
  y = abs(x)^0.73;
  if x<0
    rgbpa(i) = 1.0 - 40.0*y/(y+2.0);
  else
    rgbpa(i) = 1.0 + 40.0*y/(y+2.0);
  end
end

rgbpwa = zeros(3,1);
for i = 1:3
  x = fl*rgbpw(i)/100;
  y = abs(x)^0.73;
  if x<0
    rgbpwa(i) = 1.0 - 40.0*y/(y+2.0);
  else
    rgbpwa(i) = 1.0 + 40.0*y/(y+2.0);
  end
end

% step 6
a = rgbpa(1) - 12*rgbpa(2)/11 + rgbpa(3)/11;
b = (rgbpa(1) + rgbpa(2) - 2*rgbpa(3))/9;

% step 7
[C,h] = car2pol([a b]);
% note that C is not used
```

```
% step 8
ehH = [ 20.14 90.00 164.25 237.53 380.14
0.8 0.7 1.0 1.2 0.8
0.0 100.0 200.0 300.0 400.0 ];
hh = h;
if h<ehH(1,1)
hh = 360+ehH(1,1);
end
for k = 1:4
if hh <ehH(1,k+1)
i = k;
break;
end
end
e = ehH(2,i)+(ehH(2,i+1)-ehH(2,i))*(hh-ehH(1,i))/
(ehH(1,i+1)-ehH(1,i));
hq=ehH(3,i)+100.0*( (hh-ehH(1,i))/ehH(2,i) )/
( (hh-ehH(1,i))/ehH(2,i) + (ehH(1,i+1)-hh)/ehH(2,i+1) );
k = floor(hq/100);
Hp = floor(100.0*(hq/100-k)+0.5);
if k == 0
Hc(1).Colour = 'Yellow';
Hc(2).Colour = 'Red';
end
if k == 1
Hc(1).Colour = 'Green';
Hc(2).Colour = 'Yellow';
end
if k == 2
Hc(1).Colour = 'Blue';
Hc(2).Colour = 'Green';
end
if k == 3
Hc(1).Colour = 'Red';
Hc(2).Colour = 'Blue';
end
Hc(1).Portion = Hp;
Hc(2).Portion = 100-Hp;

cd = sprintf('%2.2f %s %2.2f %s',Hc(1).Portion, Hc(1).
Colour, Hc(2).Portion, Hc(2).Colour);
```

```
% step 9
A = (2*rgbpa(1) + rgbpa(2) + rgbpa(3)/20 - 2.05)*nbb;
Aw = (2*rgbpwa(1) + rgbpwa(2) + rgbpwa(3)/20 - 2.05)*nbb;

% step 10
z = 1 + sqrt(fll*n);
j = 100*(A/Aw)^(c*z);

% step 11
q = (1.24/c)*((j/100)^0.67)*((Aw + 3)^0.9);

% step 12
s = (5000*sqrt(a^2+b^2)*e*10*nc*nbb/13)/(rgbpa(1) + ...
rgbpa(2) + 21*rgbpa(3)/20);

% step 13
c = 2.44*s^0.69*(j/100.)^(0.67*n)*(1.64-0.29^n);

% step
14 m = c*fl^0.15;
```

An example of the function is provided by the following code segment:

```
clear

% test sample
x = 0.3618; y = 0.4483; Y = 23.93;
xyz(1) = x*Y/y; xyz(2) = Y; xyz(3) = (1-x-y)*Y/y;

% white in test light
x = 0.4476; y = 0.4074; Y = 90.00;
xyzw(1) = x*Y/y; xyzw(2) = Y; xyzw(3) = (1-x-y)*Y/y;

xyz = xyz';

xyzw = xyzw';

yb = 18.0;
la = 200;
lw = la*100/yb;
```

```
f = 1.0;
c = 0.69;
fll = 1.0;
nc = 1.0;

para = [f c fll nc];

[j,C,hq,m,h,s,q,cd] = ciecam97s(xyz,xyzw,la,yb,para)
```

The output of the *ciecam97s* function for the set of input values shown is

```
j = 34.6984
c = 39.6202
hq = 239.4437
m = 39.6202
h = 190.0318
s = 86.9153
q = 17.8702
cd = 39.00 Blue 61.00 Green
```

7

Characterization of Computer Displays

7.1 Introduction

Linear transforms are fundamental to the study of colorimetry and have many applications, especially in the characterization of imaging devices such as monitors, cameras and printers. Device calibration is concerned with setting the imaging device to a known state and ensures that the device is producing consistent results. Characterization is the relationship between device coordinates (usually *RGB* or *CMYK*) and some device-independent colour space such as CIE *XYZ* (Fairchild, 1998; Johnson, 2002). Green (2002b) argues that there are three main methods for achieving this mapping: physical models, look-up tables and numerical methods. Physical models often include terms for various properties of the device such as the absorbance, scattering and reflectance of colorants. The Kubelka–Munk model is an example of a physical model that can be used as the basis of a characterization method for a printer (Kang, 1994; Johnson, 1996). Similarly, the gain–offset–gamma model (also known as GOG) is a physical model of a computer- or visual-display unit based on a cathode-ray tube (CRT) that can be used for the characterization of most display monitors. Look-up tables define the mapping between a device space and a CIE colour space at a series of discrete measured coordinates within the colour space and may interpolate the values for intermediate coordinates. For numerical methods a series of coefficients is determined, usually based upon a set of measured samples, without prior assumptions about the physical behaviour of the device or its associated media. Examples of numerical methods include linear transforms, non-linear transforms or polynomials and artificial neural networks. A key property of any transform is whether it can easily be inverted. The advantage of a linear transform is that it is trivial to invert whereas many empirical models are not easily inverted (Iino and Berns, 1998). If inversion is not possible, then

Computational Colour Science Using MATLAB. By Stephen Westland and Caterina Ripamonti.
© 2004 John Wiley & Sons, Ltd: ISBN 0 470 84562 7

iteration may be required to perform the inverse mapping (Hardeberg, 2001). Bala (2003) provides an excellent source of further information on computational methods for device characterization.

For many devices the process of characterization can be considered to consist of two stages. The first stage performs a linearization, sometimes termed gamma correction, for certain devices. The second stage transforms the linearized values into the CIE XYZ tristimulus values. Practical device characterization will almost certainly require, in addition, that the spatial and temporal properties of the device be accounted for. Johnson (2002) notes that, even if a non-linear transform is used, usually it is better to perform the linearization process and then use approximately linear values as input to the non-linear transform.

In this chapter we describe some methods for the characterization of computer-display devices or monitors.

7.2 Gamma

The luminance generated by a computer monitor generally is not a linear function of the applied signal. Most CRT devices exhibit a power-law response to voltage so that the luminance produced at the face of the display is approximately proportional to the applied voltage raised to a power in the range 2.35–2.55 (Poynton, 2002). The value of the exponent of this power function is sometimes called the gamma of the CRT or monitor. Figure 7.1 shows the relationship between applied voltage and displayed luminance for a typical CRT at three different settings of picture control (sometimes referred to as contrast).

In a typical 8-bit digital-to-analogue converter (DAC), the lowest voltage shown in Figure 7.1 will be coded by the value 0, whereas the highest voltage will be coded by the value 255 (2^8-1).

The relationship between the voltage applied to the CRT's phosphors and the displayed luminance can be approximated by the gamma relationship

$$L = V^{\gamma}, \tag{7.1}$$

where L is the luminance of the display, V is the applied voltage (this is linearly related to the RGB values) and γ is the gamma.

7.3 The GOG model

Although all vacuum tubes, including CRTs, exhibit an inherent non-linearity, the term gamma is commonly used to represent the non-linearity of the entire opto-electronic transfer function of the display system. Berns *et al.* (1993a, 1993b) have studied the relationship between the digital monitor values (sometimes referred to as DAC values) and the displayed luminance for a range of typical

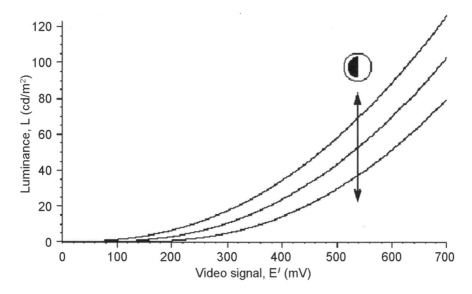

Figure 7.1 Typical transfer function of a CRT at three different settings of the picture control. Reproduced from Poynton (2002) with permission from the author

CRT devices. Equation (7.2) describes a realistic relationship between the luminance L_r and DAC value d_r for the red channel (Berns and Katoh, 2002),

$$L_r = k_{\lambda r}[a_r[(v_{max} - v_{min})(\text{LUT}(d_r)/(2^N - 1)) + v_{min}] + b_r - v_{cr}]^{\gamma r} \qquad (7.2)$$

where LUT is a function that represents the video look-up table, N is the number of bits in the DAC, v_{min} and v_{max} are the minimum and maximum voltages of the video-signal generator, a_r and b_r are the CRT video amplifier gain and offset, v_{cr} is the cut-off voltage defining zero beam current, λr is the gamma of the channel and $k_{\lambda r}$ is a spectral constant accounting for the particular CRT phosphors and faceplate combination. In addition, $L_r = 0$ if $v_{cr} \leqslant a_r$ $[(v_{max} - v_{min})(\text{LUT}(d_r)/(2^N - 1)) + v_{min}] + b_r$. Similar relationships can be expressed for the blue and green channels in a display device.

Generally, an accurate physical model of monitor behaviour is not used for the purposes of characterization. Rather, the relationship between luminance L and DAC $d/(2^N - 1)$ is generalized to yield

$$L = (ad/(2^N - 1) + b)^\gamma, \qquad (7.3)$$

where it can be useful to think of the coefficients a and b as the system gain and offset, respectively. This generalized relationship is known as the gain–offset–gamma (GOG) model (Berns and Katoh, 2002). The implication of this equation is that although the CRT has an inherent fixed gamma, the effective gamma of a system will be dependent upon how the offset and gain controls are set. In practice, a relationship of the form shown as Equation (7.3) is used to map the normalized DAC values ($d_r/(2^N-1)$, $d_g/(2^N-1)$, and $d_b/(2^N-1)$) to the linearized normalized DAC values (R, G, and B). Thus, for the red channel the following equation can be used,

$$R = (ad_r/(2^N - 1) + (1 - a))^\gamma, \qquad (7.4)$$

where the normalization procedure requires that the system gain and offset are equal to unity. Since there are three model parameters but only two degrees of freedom, a minimum of two radiometric measurements are required per channel. The advantage of minimizing the number of measurements required to characterize the monitor is important since it is widely recognized that when making the measurements a time of at least 80 s must be allowed for the colour to stabilize (Berns *et al.*, 1993a, 1993b). It is only practicable to allow this time for relatively small numbers of measurements. Berns *et al.* (1993a) recommend measuring neutral colours where the load is placed equally across all three channels rather than highly chromatic colours where the load is placed on only one of the gun amplifiers. As few as two neutral colours need be measured in order to be able to determine the parameters of Equation (7.4) for all three channels.

7.4   Device-independent transformation

Once the GOG model [Equation (7.4)] has been used to linearize the DAC values, the values can be related to tristimulus values using a simple linear transform,

$$\begin{bmatrix} X \\ Y \\ Z \end{bmatrix} = \begin{bmatrix} X_{r,\max} & X_{g,\max} & X_{b,\max} \\ Y_{r,\max} & Y_{g,\max} & Y_{b,\max} \\ Z_{r,\max} & Z_{g,\max} & Z_{b,\max} \end{bmatrix} \begin{bmatrix} R \\ G \\ B \end{bmatrix}, \qquad (7.5)$$

where RGB are the linearized and normalized (in the range 0–1) DAC values. Three measurements are required in order to specify the system matrix for Equation (7.5). The tristimulus values XYZ must be measured for each of the guns at the maximum DAC value (2^N-1, where N is the number of bits in the DAC). The XYZ values of the red gun at maximum intensity form the first column of the system matrix [Equation (7.5)] and the XYZ values for the green and blue guns form the second and third columns, respectively. Once each of the guns has been measured for maximum DAC values the system matrix for Equation (7.5) is known.

7.5 Typical characterization procedure

Before making radiometric measurements for characterization the monitor should be placed in the position where it will be used and then turned on (for most monitors a deguassing process takes place whenever the monitor is initially powered up). Sufficient time should be allowed for the monitor to warm up. The warm-up time required for a monitor to stabilize after initial power-up varies for different devices but can range from 15 min to 3 h or more (Berns *et al.*, 1993b). For accurate characterization it is important that the monitor exhibits good spatial independence and channel independence. Spatial independence can be assessed by measuring the colour difference between a white patch displayed in the centre of the screen with a black surround and a physically identical white patch with a white surround. Berns *et al.* (1993b) measured spatial independence using this technique for five different monitors and found CIELAB colour differences between 2.6 (for the best monitor) and 17.4 (for the worst monitor). Channel independence can be assessed by computing the colour difference between full-field white and the prediction of the full-field white obtained by adding the tristimulus values of the full-field pure red, green and blue conditions. Berns *et al.* found that the channel-independence error can be minimized by reducing the maximum value of luminance that can be displayed. Experience suggests that for many monitors a maximum display luminance of about 80 cd/m^2 provides a suitable compromise between being able to achieve good characterization and being able to display reasonable brightness levels.

For characterization purposes it is recommended that patches be displayed at the centre of the monitor against a neutral field set at about one-fifth of the luminance of the maximum brightness in order that the measurements are taken in typical conditions. A spectroradiometer or spectrocolorimeter should be used to measure the luminance and chromaticities of each of the calibration patches [note that Equation (4.10) can be used to recover the tristimulus values from the luminance and two of the chromaticity coordinates]. Three measurements are needed to obtain the maximum tristimulus values of each of the guns. The digital input values [d_r d_g d_b] for these patches for a system with 8 bits per channel should be [255 0 0], [0 255 0] and [0 0 255]. These three measurements should be used to define the system matrix for Equation (7.5).

Measurements of as few as two neutral patches are then made in order to allow the parameters of the GOG model to be computed but in practice normally about five neutral patches are used (Luo, 2003). The tristimulus values of the neutral samples are measured using a spectroradiometer and then Equation (7.5) is inverted to predict the linearized normalized DAC values *RGB*. For each of the neutral samples and for each channel the normalized DAC values and the linearized normalized DAC values are then known, and therefore the GOG parameters may be determined using a multidimensional optimization technique such as the simplex algorithm.

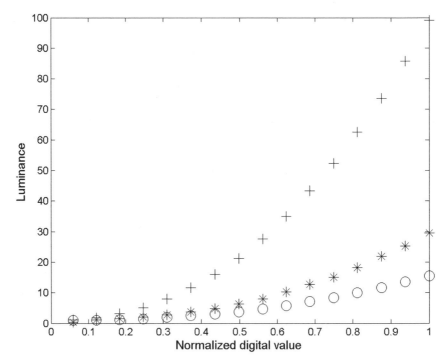

Figure 7.2 Measured luminance response curve for normalized R (∗), G (+) and B (○) digital gun values

Whereas CRT devices exhibit a power-law relationship between the DAC values and the output luminance, liquid crystal display (LCD) devices often exhibit an electro-optic response that is better modelled as a sigmoidal or S-shaped function (Sharma, 2002). Many LCD manufacturers, however, build correction tables into the video card that result in the LCD response mimicking the response of a CRT with a power law of about 2.0 (Bala, 2003).

7.6 Implementations and examples

Measurements were made (Owens, 2002a) for a typical monitor and are used in this section to illustrate the implementation of a typical characterization procedure. Figure 7.2 shows the luminance values measured for a range of different monitor values for each channel separately; the typical non-linear response is evident.

The brightness setting of the monitor was adjusted so that the white was approximately $90\,\text{cd/m}^2$. The tristimulus values of each of the channels at maximum output were then measured to allow the linear transform from linearized normalized DAC values to be written as

Table 7.1 Measured data for the monitor characterization example

d_r	d_g	d_b	X	Y	Z
255	0	0	29.67	16.30	2.12
0	255	0	26.62	54.90	10.28
0	0	255	17.78	8.48	91.51
40	40	40	2.25	2.42	2.94
90	90	90	8.26	8.95	11.38
140	140	140	19.84	21.50	27.79
190	190	190	37.93	41.10	53.13
240	240	240	63.23	68.30	88.79

$$\begin{bmatrix} X \\ Y \\ Z \end{bmatrix} = \begin{bmatrix} 29.67 & 26.62 & 17.78 \\ 16.30 & 54.90 & 8.48 \\ 2.12 & 10.28 & 91.51 \end{bmatrix} \begin{bmatrix} R \\ G \\ B \end{bmatrix}. \tag{7.6}$$

Five neutral measurements were made to allow the GOG parameters to be estimated. Table 7.1 lists the details of the eight measurements in total that were used for the characterization of the monitor.

Seven test samples were also measured (not shown in Table 7.1) for testing the characterization model. The following MATLAB code illustrates how the data in Table 7.1 can be used to characterize the display using the functions *testgog* and *compgog*, which are then described.

```
clear

% define the dac RGB and XYZ values of the known samples
r = [255 0 0 29.67 16.30 2.12] ; % R = 255 ; G = 0 ; B = 0
g = [0 255 0 26.62 54.90 10.28] ; % R = 0 ; G = 255 ; B = 0
b = [0 0 255 17.78 8.48 91.51] ; % R = 0 ; G = 0 ; B = 255
% define the dac RGB and XYZ values of the neutral samples
% each row contains R G B X Y Z
N = [40 40 40 2.25 2.42 2.94; 90 90 90 8.26 8.95 11.38; ...
    140 140 140 19.84 21.50 27.79; 190 190 190 37.93 ...
    41.10 53.13; 240 240 240 63.23 68.30 88.79];
% define the dac RGB and XYZ values of the test samples
T = [200 200 100 34.62 42.20 19.09; 100 100 200 17.80 ...
    14.50 54.53; 200 100 50 21.46 17.30 5.95; 25 150 250 ...
    25.58 25.30 90.46; 250 50 25 29.73 17.80 3.47; 175 ...
    250 50 38.51 59.50 13.62; 25 10 50 1.65 1.28 3.81];

% specify the matrix A to convert RGB to XYZ
```

```
A = [r(4) g(4) b(4) ; r(5) g(5) b(5) ; r(6) g(6) b(6)];

% compute the matrix AI to convert XYZ to RGB
AI = inv(A);

% obtain the XYZ values of the neutral patches
NXYZ = N(:,4:6);

% compute the RGB values of the neutral patches
NRGB = (AI*NXYZ')';

% obtain the normalised dac values of the neutral patches
DRGB = N(:,1:3)/255;

% compute the GOG values for each channel
x1 = linspace(0,1,10);

x = [1, 1];
options = optimset;
x = fminsearch('gogtest',x,options,DRGB(:,1),NRGB(:,1));
gogvals(1,:) = x;
figure
plot(DRGB(:,1),NRGB(:,1),'r*')
y1 = compgog(gogvals(1,:),x1);
hold on
plot(x1,y1,'r-')

x = [1, 1];
options = optimset;
x = fminsearch('gogtest',x,options,DRGB(:,2),NRGB(:,2));
gogvals(2,:) = x;
hold on
plot(DRGB(:,2),NRGB(:,2),'g*')
y1 = compgog(gogvals(2,:),x1);
hold on
plot(x1,y1,'g-')

x = [1, 1];
options = optimset;
options.TolFun = 0.0000001;
x = fminsearch('gogtest',x,options,DRGB(:,3),NRGB(:,3));
gogvals(3,:) = x;
hold on
```

```
plot(DRGB(:,3),NRGB(:,3),'b*')
y1 = compgog(gogvals(3,:),x1);
hold on
plot(x1,y1,'b-')

disp('gog values')
disp(gogvals)

RGB = zeros(3+length(N(:,1)),3);
RDACS = [R(1); G(1); B(1); N(:,1)]/255;
RGB(:,1) = compgog(gogvals(1,:), RDACS);
RDACS = [R(2); G(2); B(2); N(:,2)]/255;
RGB(:,2) = compgog(gogvals(2,:), RDACS);
RDACS = [R(3); G(3); B(3); N(:,3)]/255;
RGB(:,3) = compgog(gogvals(3,:), RDACS);

XYZ = (A*RGB')';

AXYZ = [R(4:6); G(4:6); B(4:6); N(:,4:6)];

for i = 1:8
  [lab1] = xyz2lab(XYZ(i,:),'d65_64');
  [lab2] = xyz2lab(AXYZ(i,:),'d65_64');
  [thisDE] = cielabde(lab1,lab2);
  de(i) = thisDE;
end

disp ('known values')
disp(de)

RGB = zeros(length(T(:,1)),3);
RDACS = [T(:,1)]/255;
RGB(:,1) = compgog(gogvals(1,:), RDACS);
RDACS = [T(:,2)]/255;
RGB(:,2) = compgog(gogvals(2,:), RDACS);
RDACS = [T(:,3)]/255;
RGB(:,3) = compgog(gogvals(3,:), RDACS);

XYZ = (A*RGB')';

AXYZ = [T(:,4:6)];
clear de
```

```
% now compute the error for the test samples
for i = 1:7
  [lab1] = xyz2lab(XYZ(i,:),'d65_64');
  [lab2] = xyz2lab(AXYZ(i,:),'d65_64');
  [thisDE] = cielabde(lab1,lab2);
  de(i) = thisDE;
end

disp('test values')
disp(de)
```

In the preceding code the matrices **r**, **g** and **b** contain the DAC *RGB* values and the measured *XYZ* values for the pure channel colours. The matrix **N** contains the data for the five neutral samples (see Table 7.1) and the matrix **T** contains data for the seven test samples. The function *gogtest* computes the root-mean-squared error between actual DAC values and the predicted DAC values. The required inputs to the *gogtest* function are **gogs** (a 2×1 column matrix containing the gamma and gain terms, respectively), **dacs** (an $n \times 1$ column matrix of normalized *RGB* values) and **rgbs** (an $n \times 1$ column matrix of predicted *RGB* values). The predicted *RGB* values are obtained by inverting Equation (7.6) and using the measured *XYZ* values. The format of the call to *gogtest* is

```
[err] = gogtest(gogs,dacs,rgbs)
```

where the returned 1×1 matrix **err** contains the error using the GOG parameters defined by **gogs**. This function is useful since it allows **gogs** to be optimized for the minimum value of **err** using a suitable optimization method.

Box 18: *gogtest.m*

```
function [err] = (gogs,dacs,rgbs)

% function [err] = gogtest(gogs,dacs,rgbs)
% computes the error between measured and predicted
% linearized dac values for a given set of GOG values
% gogs is a 2 by 1 matrix that contains the gamma and gain
% dacs is an n by 1 matrix that contains the actual RGB values
```

```
% obtained by dividing the RGB values by 255
% rgbs is an n by 1 matrix that is obtained from a linear
% transform of measured XYZ values
gamma = gogs(1);
gain = gogs(2);

% force to be row matrices
dacs = dacs(:)';
rgbs = rgbs(:)';

if (length(dacs) ~= length(rgbs))
  disp('dacs and rgbs vectors must be the same length');
  err = 0;
  return
end

% compute gog model predictions
for i = 1:length(dacs)
  if (gain*dacs(i) + (1-gain)) <= 0
    pred(i) = 0;
  else
    pred(i) = (gain*dacs(i) + (1-gain))^gamma;
  end
end

% force to be a row matrix
pred = pred(:)';
% compute rms error
err = sqrt((sum((rgbs-pred).*(rgbs-pred)))/...
length(dacs));
```

The MATLAB function *fminsearch* performs a multidimensional uncon-strained non-linear minimization. Once the parameters have been determined, the function *compgog* can be used to implement the GOG model. The format of the call to *compgog* is

```
[rgb] = compgog(gogs,dacs)
```

where the returned $n \times 1$ matrix **rgb** contains linearized R, G or B values for the n samples defined by **dacs**.

Box 19: *compgog.m*

```
function [rgb] = compgog(gogs,dacs)

% function [rgb] = compgog(gogs,dacs)
% computes the linearized RGB values
% from the normalized RGB values
% for a given set of gog values
% gog is a 2 by 1 matrix that contains the gamma and gain
% dacs is an n by 1 matrix that contains the RGB values
% rgb is an n by 1 matrix of linearized RGB values

gamma = gogs(1);
gain = gogs(2);
for i = 1:length(dacs)
  if (gain*dacs(i) + (1-gain)) <= 0
    rgb(i) = 0;
  else
    rgb(i) = (gain*dacs(i) + (1-gain))^gamma;
  end
end
% force output to be a column vector
rgb = rgb(:);
```

The characterization performance for the eight stimuli defined in Table 7.1 and the additional seven test samples (these were not used in the characterization procedure) is, on average, just under 1.7 CIELAB units.

An additional function, *rgb2xyz*, has been provided to convert *RGB* DAC values (in the range 0–255) to CIE *XYZ* values directly. The format of the *rgb2xyz* function is

```
[xyz] = rgb2xyz(dacs, gogvals, A),
```

where **dacs** is a 3×1 matrix of *RGB* DAC values and the returned 3×1 matrix **xyz** contains the predicted CIE values. The 3×2 matrix **gogvals** must contain the gamma and gain terms of the GOG model for each of the three channels, whereas the matrix **A** must contain the 3×3 matrix that transforms linearized *RGB* values to *XYZ* values.

Box 20: *rgb2xyz.m*

```
function [XYZ] = rgb2xyz(dacs, gogs, A)

% function [XYZ] = rgb2xyz(dacs, gogs, A)
% converts RGB DACS from a monitor to CIE XYZ
% dacs is a 3 by 1 matrix containing the RGB DACS (0-255)
% gogs is a 2 by 1 matrix containing the gamma and gain
% A is a 3 by 3 matrix to transform RGB to XYZ

dacs = dacs(:)'; % force to be a row matrix
if (length(dacs) ~= 3)
  disp('DACS must be 3 by 1 or 1 by 3'); return;
end

dacs = dacs/255;

RGB(1) = compgog(gogs(1,:), dacs(1));
RGB(2) = compgog(gogs(2,:), dacs(2));
RGB(3) = compgog(gogs(3,:), dacs(3));

RGB = RGB(:);

XYZ = A*RGB;
```

The inverse function *xyz2rgb* is also provided.

Box 21: *xyz2rgb.m*

```
function [dacs] = xyz2rgb(XYZ)

% function [dacs] = xyz2rgb(XYZ, gogvals, A)
% converts XYZ to RGB DACS for a monitor
% XYZ is a 3 by 1 matrix containing the XYZ values
% gogvals is a 3 by 2 matrix containing the gamma and gain
% for each of the three channels
% A is a 3 by 3 matrix to transform RGB to XYZ
```

```
RGB = inv(A)*XYZ;

dacs(1) = compigog(gogvals(1,:), RGB(1));
dacs(2) = compigog(gogvals(2,:), RGB(2));
dacs(3) = compigog(gogvals(3,:), RGB(3));

dacs = dacs*255;

if (dacs(1)>255)
    dacs(1) = 255;
end
if (dacs(2)>255)
    dacs(2) = 255;
end
if (dacs(3)>255)
    dacs(3) = 255;
end

if (dacs(1)<0)
    dacs(1) = 0;
end
if (dacs(2)<0)
    dacs(2) = 0;
end
if (dacs(3)<0)
    dacs(3) = 0;
end

dacs = dacs(:);
```

The function *xyz2rgb* requires an inverse version of the *compgog* function and this is provided with the *compigog* function.

Box 22: *compigogs.m*

```
function [dacs] = compigog(gogs,rgb)

% function [dacs] = compgog(gogs,rgb)
% computes the normalized RGB values
```

```
% from the linearized RGB values
% for a given set of gog values
% gog is a 2 by 1 matrix that contains the gamma and gain
% dacs is an n by 1 matrix that contains the RGB values
% rgb is an n by 1 matrix of linearized RGB values

gamma = gogs(1);
gain = gogs(2);
for i = 1:length(rgb)
  dacs(i) = ((rgb(i)^(1/gamma)) - (1-gain))/gain;
end
% force output to be a column vector
rgb = rgb(:);
```

8

Characterization of Cameras

8.1 Introduction

Some general comments regarding characterization can be found in Chapter 7, Section 7.1. For input devices such as scanners and cameras it is important to note that effective characterization is only practicable if the device does not perform automatic white-point balancing. Automatic white-point balancing is where the *RGB* values of each pixel in the captured image are transformed so that the pixel for the brightest patch in the image scene is denoted as white with equal *RGB* values (normally $R = G = B = 255$). White-point balancing can be useful if the aim is to capture a pleasing image, since the human visual system is able to discount the colour of the light source so that surfaces tend to retain their daylight appearance. For colorimetric characterization, however, this setting should be disabled if at all possible.

The most efficient method for characterizing a digital camera or scanner is to image a chart containing a set of colours of known tristimulus values (Johnson, 2002). Such charts commonly include neutral patches that may be used to linearize the camera *RGB* outputs and coloured patches that may be used to characterize a transform from linearized *RGB* values to CIE *XYZ* values. In the late 1980s a working group of the ANSI IT8 (Image Technology Committee No. 8) was created to define standard targets to be used in the characterization of scanners and printers (McDowell, 2002). The IT8 committee chose to colorimetrically define the colours that should appear in the target, but then allow individual manufacturers to produce targets to meet these requirements. Two standards were developed, ANSI IT8.7/1 and ANSI IT8.7/2, for transmission and reflectance modes, respectively, and they were combined into a single ISO standard (ISO 12641:1997). Two further charts that are sometimes used for device characterization are the Macbeth ColorChecker chart (which contains 24 patches) and the Macbeth ColorChecker DC chart (which contains over 200 patches).

Computational Colour Science Using MATLAB. By Stephen Westland and Caterina Ripamonti.
© 2004 John Wiley & Sons, Ltd: ISBN 0 470 84562 7

8.2   Correction for non-linearity

Although the response of charge-coupled diode (CCD) material is approximately linearly related to the intensity of the light falling on it, it is unlikely that the *RGB* outputs of a scanner or digital camera will be linearly related to the *XYZ* tristimulus values of the surfaces in the scene. The raw channel responses are invariably processed by on-board software in an attempt to generate *RGB* responses that are more closely matched to the colour-matching functions than are allowed by current methods for producing filters. Typically, the raw *RGB* values may be transformed by a 3×3 linear matrix to give the output *RGB* values. Furthermore, many manufacturers impose a non-linearity during this 'matrix-mixing' stage to approximately match the inverse of the non-linearity of display systems or as part of the solution to provide high signal-to-noise ratios. It is therefore common to consider a correction for non-linearity as the first stage of a camera- or scanner-characterization process. For a digital camera we may, for example, consider a relation of the form

$$C_c = (C'_c)^p, \tag{8.1}$$

where C'_c is the raw response of the camera channel c, p is an exponent for the channel, and C_c is a transformed camera response for that channel that is linear to the channel input. A set of grey-scale samples is often used to empirically determine the exponent p. Thus, the raw camera responses are determined for a range of grey samples under a constant and known light source. The *XYZ* values then can easily be computed for the grey samples and linearization is achieved by finding the value of the exponent p such that there is a linear relationship between C_c and Y for the set of grey samples.

However, in order to correctly determine p [according to Equation (8.1)] we need to know the values of C_c and C'_c for each of the grey patches. The value of C'_c is the input to the channel and the value of C_c is the channel response following non-linear processing (Thomson and Westland, 2002). The input to the channel may easily be computed if the spectral reflectance of the sample, the spectral power of the illumination and the channel sensitivity are all known. Unfortunately, it is usually the case that the channel sensitivity is not known. If p is determined so that there is a linear relationship between C_c and Y, then it is clear that the Y colour-matching function or photopic sensitivity is being used as a crude estimate of the channel sensitivity. For grey samples imaged under an equal-energy illuminant, Y and C'_c are approximately linearly related but this proportionality decreases for increasingly chromatic samples and light sources. It is interesting to note that if the channel sensitivity was known, and the true values of C'_c could be computed, then it would not be necessary to use grey samples to linearize the channel responses; any samples could be used. Since the spectral sensitivities generally are not known, however, linearization is typically established using achromatic samples. The grey samples of the Macbeth

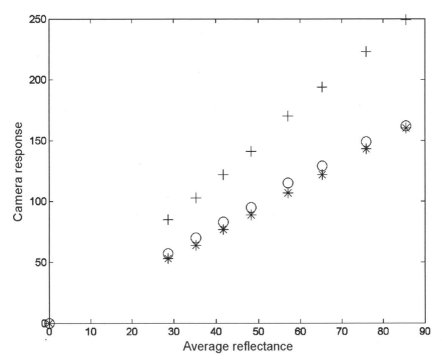

Figure 8.1 Measured camera response for red (∗), green (+) and blue (○) channels for neutral ColorChecker patches of known reflectance

ColorChecker provide a convenient grey scale for this purpose. Some workers choose to linearize the camera responses with the appropriate tristimulus values (X for the red channel, Y for the green channel and Z for the blue channel) or with the average reflectance of the samples.

Figure 8.1 illustrates the relationship between the RGB values (C_R, C_G and C_B) measured using a typical good-quality colour camera and the average spectral reflectance values for the achromatic samples of the Macbeth ColorChecker. Note that for a good-quality camera the camera responses often exhibit an approximately linear relationship with the mean reflectance or Y tristimulus value of the grey samples.

8.3 Device-independent representation

If a linear relationship exists between the device's channel outputs and the tristimulus values, then it is possible to determine the transform to XYZ from as few as three known samples. Thus, if we represent the n known XYZ values by the $3 \times n$ matrix **T** and the n recorded RGB values by the $3 \times n$ matrix **C**, then

$$\mathbf{T} = \mathbf{AC}, \tag{8.2}$$

where **A** is the 3×3 system matrix. If three suitable samples are available, then the linear system is exactly determined. Further samples could be used, to over-determine the system, but are only strictly necessary if a linear transform does not exist between the two colour spaces. In this situation it is usually preferable to use a non-linear transform. Johnson (2002) notes that, even if a non-linear transform is used, it is usually better to perform a linearization process and then use approximately linear values as input to the non-linear transform.

Various non-linear transforms can be used such as

$$
\begin{aligned}
X &= a_{11}R + a_{12}G + a_{13}B + a_{14}R^2 + a_{15}G^2 + a_{16}B^2 + a_{17}RGB + a_{18}, \\
Y &= a_{21}R + a_{22}G + a_{23}B + a_{24}R^2 + a_{25}G^2 + a_{26}B^2 + a_{27}RGB + a_{28}, \quad (8.3) \\
Z &= a_{31}R + a_{32}G + a_{33}B + a_{34}R^2 + a_{35}G^2 + a_{36}B^2 + a_{37}RGB + a_{38},
\end{aligned}
$$

where, in this case, a total of 24 coefficients need to be determined. In matrix notation we can write

$$\mathbf{T} = \mathbf{AD}, \tag{8.4}$$

where the system matrix **A** is now a 3×8 matrix of the coefficients a_{11}–a_{38}. The matrix **D** is the $8 \times n$ column matrix of augmented device responses [in the case of Equation (8.3) this is given by the terms R, G, B, R^2, G^2, B^2, RGB and 1].

The system is determined by computing the pseudoinverse (see Chapter 2, Section 2.4) of the augmented matrix; thus

$$\mathbf{A} = \mathbf{D}^+\mathbf{T}. \tag{8.5}$$

As an alternative to linear or non-linear transforms of this type it is also possible to use a neural network to perform a mapping from **C** to **T**. However, it has been shown that neural networks offer no advantage over polynomial transforms for camera characterization (Cheung and Westland, 2002) and yet can be difficult and time-consuming to train.

8.4   Implementations and examples

The first stage in characterizing an input device such as a scanner or a camera is to linearize the measured RGB values. Table 8.1 lists the camera RGB values and the mean reflectances for the grey samples of the Macbeth ColorChecker chart which were measured using a typical high-end digital camera (Cheung and Westland, 2002). Note that the first row of data in Table 8.1 does not show measured values but implies that the camera gives a zero response for a zero signal.

Figure 8.1 shows the relationship between the camera responses and the mean reflectance P for the neutral patches of the ColorChecker. It is noticeable that there is an approximately linear relationship between the RGB values and the P

Table 8.1 Measurements used for the camera characterization. The first column gives the mean reflectance for the neutral samples of the Macbeth ColorChecker and the second to fourth columns give the recorded camera responses

P	R	G	B
0	0	0	0
28.71	53	85	57
35.20	64	103	70
41.71	77	122	83
48.30	89	141	95
57.09	107	170	115
65.31	122	194	129
75.88	143	223	149
85.44	160	249	162

values; this is quite typical for a high-end camera. For low-end cameras the relationship often is very non-linear. Linearization may be achieved by plotting the camera response for each channel and then fitting the data with a low-order polynomial. The fitted polynomial is then used to transform the raw camera responses to linear camera responses.

A function called *getlincam* has been written to perform the polynomial fitting for the three channels.

Box 23: *getlincam.m*

```
function [out] = getlincam(p,RGB,graphs)

% function [out] = getlincam(p,RGB,graphs)
% function to compute polynomial fits for camera
% grey-scale data. The inputs are p (a set of n by
% 1 mean reflectance values) and RGB ( a set of 3 by
% n RGB triplets). If graphs is set to 'on' then
% a plot of the fits is generated

r = RGB(1,:)/255;
g = RGB(2,:)/255;
b = RGB(3,:)/255;
ref = p/100;

if nargin<3
  plotgraphs = 0;
```

```
else
  plotgraphs = strcmp('on',graphs);
end

% fit the low-order polynomials
[pr1,sr1]=polyfit(r,ref,3);
[pg1,sg1]=polyfit(g,ref,3);
[pb1,sb1]=polyfit(b,ref,3);

% plot graphs if the plotgraphs variable is set
if (plotgraphs)
  figure

  subplot(3,2,1)
  plot(r,ref,'ko');
  [pr1,sr1]=polyfit(r,ref,3);
  x = linspace(0,1,11);
  y = polyval(pr1,x);
  hold on
  plot(x,y,'k-');
  ylabel('Y')
  xlabel('R channel');
  axis([0 1 0 1])
  subplot(3,2,2)
  py = polyval(pr1,r);
  plot(ref,py,'ko');
  hold on
  plot([0 1], [0 1], 'k-');
  axis([0 1 0 1])
  disp(255*py')

  subplot(3,2,3)
  plot(g,ref,'ko');
  x = linspace(0,1,11);
  y = polyval(pg1,x);
  hold on
  plot(x,y,'k-');
  ylabel('Y')
  xlabel('G channel');
  axis([0 1 0 1])
  subplot(3,2,4)
  py = polyval(pg1,g);
```

```
plot(ref,py,'ko');
hold on
plot([0 1],[0 1],'k-');
axis([0 1 0 1])
disp(255*py')

subplot(3,2,5)
plot(b,ref,'ko');
x = linspace(0,1,11);
y = polyval(pb1,x);
hold on
plot(x,y,'k-');
ylabel('Y')
xlabel('B channel');
axis([0 1 0 1])
subplot(3,2,6)
py = polyval(pb1,b);
plot(ref,py,'bo');
hold on
plot([0 1],[0 1],'k-');
axis([0 1 0 1])
   disp(255*py')

end

out = [pr1; pg1; pb1];
```

The format for this function is

```
[CALDATA] = getlincam(p, RGB, graphs)
```

where **p** is an $n \times 1$ matrix containing the mean reflectance of the n neutral patches and **RGB** is a $3 \times n$ matrix containing the corresponding *RGB* values. Note that the first step in *getlincam* is to normalize the camera data for each of the three channels to be in the range [0, 255] and therefore there is an assumption that the data are in 8-bit-per-channel format.

In the *getlincam* script the built-in MATLAB function *polyfit* is used to fit a third-order polynomial for each channel. The output of the function is a 3×4 matrix **CALDATA**, each row of which contains the polynomial coefficients for one of the channels. An optional input argument `graphs` can be set to 'on' or

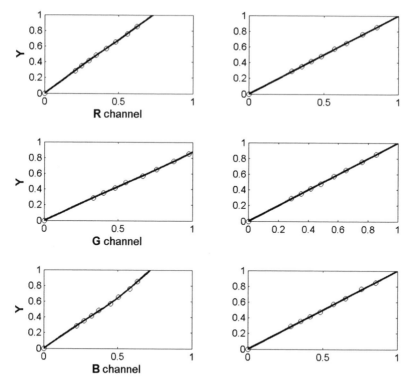

Figure 8.2 The left-hand column shows the plot of value vs. channel response (○) for the grey patches (Table 8.1) and the polynomial fits (—). The right-hand column shows the transformed camera responses plotted against value (○) and, for comparison, the ideal linear responses (—)

'off' to generate or suppress plots of the polynomial fits for visual evaluation of the goodness of the linearization. The default value of **graphs** is 'off'.

Figure 8.2 gives an example of the graphical output from *getlincam* using the data in Table 8.1 to fill the **p** and **RGB** matrices. The right-hand column of Figure 8.2 shows the transformed *RGB* data plotted against the mean reflectance for each of the neutral samples. Linear relationships are observed for each of the three channels.

A further function, *lincam*, has been written which uses the polynomial fits obtained from *getlincam* to convert raw *RGB* values into linearized *RGB* values.

Box 24: *lincam.m*

```
function [RGBout] = lincam(caldata, RGB)

% function [RGBout] = lincam(caldata,RGB)
% computes linearized camera values using
```

```
% polynomial transforms obtained from getlincam
% caldata is a 3 by 4 matrix produced from getlincam.m
% RGB is an n by 3 matrix of RGB values (in range 0-255)

RGB = RGB/255;
imsize = size(RGB);
RGB = reshape(RGB, prod(imsize)/3, 3);

for i=1:3
  x = polyval(caldata(i,:),RGB(:,i));
  RGB(:,i) = x*255;
end

RGB(RGB0) = 0;
RGB(RGB255) = 255;

RGBout = reshape(RGB,imsize);
```

The format for *lincam* is

```
[tRGB] = lincam(CALDATA,RGB)
```

where **CALDATA** is a 3×4 matrix obtained from *getlincam* and RGB is an $m \times n \times 3$ matrix of camera *RGB* values where the image size is $m \times n$. The output of the function is an $m \times n \times 3$ matrix of linearized camera *RGB* values. Unlike many of the functions in the earlier part of this book, this function operates on the whole image with a single function call. In this respect the MATLAB function *reshape* is extremely useful.

An image may be read into MATLAB using the simple instruction

```
image = imread('test.tif');
```

where `test.tif` is the name of the image file. For an *RGB* image of size 127×101, for example, the variable image would now be defined by

```
image   127 × 101 × 3   38481 uint8 array
```

It is necessary to use the double command, thus

```
image = double(image);
```

in order to convert the variable image from UINT format to DOUBLE format.

In the code for *lincam* (Box 24) the data are normalized (assuming 8 bits per channel) and then the size of the image is obtained. The *reshape* command is then used to convert, for the example, the $127 \times 101 \times 3$ data into a flatter 12827×3 format. Each column of the new format holds the data for one of the *R*, *G* or *B* channels. Following the polynomial transform, the data can be reshaped back to the original $127 \times 101 \times 3$ format. If a point-operation was applied to an image and the *R*, *G* and *B* channels could be treated as the same, then the *reshape* command would be used to generate a 38532×1 structure. In *lincam*, however, the polynomial transform is different for each of the three channels.

An interesting feature of the *getlincam* and *lincam* functions is that the data from each of the three channels are balanced for the neutral samples and this is evident in Figure 8.2. In Table 8.1, for example, the *RGB* values of the brightest neutral patch can be seen to be [160, 249, 162]. This is not ideal because such a patch would not appear to be neutral if it was displayed on a typical monitor. The transformed *RGB* values of the white patch output from *lincam*, however, are [217, 218, 217]. This process may be described as channel balancing, where the original imbalance results from the combined properties of the camera and the spectral power distribution of the light under which the images are taken. For certain applications, channel balancing may be a desirable property of the linearization function implemented by *getlincam* although, for other applications, it would be possible to modify the function so that channel balancing did not occur. The concept of channel balancing should not be confused with that of white-point balancing. The latter is a property of a great number of commercially available digital cameras and means that the brightest point in any captured image is converted to *RGB* values of [255, 255, 255]. This property is very undesirable if accurate characterization or even calibration is required since the *RGB* values obtained for a given patch may vary depending upon the properties of other patches in the image. If a digital camera performs white-point balancing, then this property must be disabled before attempting device characterization. Note that it is not desirable that the maximum values of the channel outputs are only 160 and 162 for the red and blue channels (Table 8.1) since the full 8-bit resolution of the device is not being exploited. In general, if the gain settings of the individual channels can be adjusted, then this should be set so that for a very white patch the channels give responses close to (but not equal to) 255.

Cheung and Westland (2004) have conducted a study of camera characterization and specifically have compared the ability of polynomials and neural networks to carry out this task (neural networks are described in more detail in Chapter 9). In this study using the Agfa StudioCam camera 192 samples from the centre of the Macbeth DC ColorChecker were used as training samples and the 24 samples of the Macbeth ColorChecker were used as test samples. The training samples were used to train the neural network and to determine the coefficients

Table 8.2 Polynomial models used in the camera characterization study by Cheung and Westland (2004)

$m \times 3$	Augmented matrix
3×3	$[R \quad G \quad B]$
5×3	$[R \quad G \quad B \quad RGB \quad 1]$
7×3	$[R \quad G \quad B \quad RG \quad RB \quad GB \quad 1]$
8×3	$[R \quad G \quad B \quad RG \quad RB \quad GB \quad RGB \quad 1]$
10×3	$[R \quad G \quad B \quad RG \quad RB \quad GB \quad R^2 \quad G^2 \quad B^2 \quad 1]$
11×3	$[R \quad G \quad B \quad RG \quad RB \quad GB \quad R^2 \, G^2 \quad B^2 \quad RGB \quad 1]$
14×3	$[R \quad G \quad B \quad RG \quad RB \quad GB \quad R^2 \quad G^2 \quad B^2 \quad RGB \quad R^3 \quad G^3 \quad B^3 \quad 1]$
16×3	$[R \quad G \quad B \quad RG \quad RB \quad GB \quad R^2 \quad G^2 \quad B^2 \quad RGB \quad R^2G \quad G^2B \quad B^2R \quad R^3 \quad G^3 \quad B^3]$
17×3	$[R \quad G \quad B \quad RG \quad RB \quad GB \quad R^2 \quad G^2 \quad B^2 \quad RGB \quad R^2G \quad G^2B \quad B^2R \quad R^3 \quad G^3 \quad B^3 \quad 1]$
19×3	$[R \quad G \quad B \quad RG \quad RB \quad GB \quad R^2 \quad G^2 \quad B^2 \quad RGB \quad R^2G \quad G^2B \quad B^2R \quad R^2B \quad G^2R \quad B^2G \quad R^3 \quad G^3 \quad B^3]$
20×3	$[R \quad G \quad B \quad RG \quad RB \quad GB \quad R^2 \quad G^2 \quad B^2 \quad RGB \quad R^2G \quad G^2B \quad B^2R \quad R^2B \quad G^2R \quad B^2G \quad R^3 \quad G^3 \quad B^3 \quad 1]$
22×3	$[R \quad G \quad B \quad RG \quad RB \quad GB \quad R^2 \quad G^2 \quad B^2 \quad RGB \quad R^2G \quad G^2B \quad B^2R \quad R^2B \quad G^2R \quad B^2G \quad R^3 \quad G^3 \quad B^3 \quad R^2GB \quad RG^2B \quad RGB^2]$

of the polynomial transform. The camera *RGB* values were linearized and corrected for spatial non-uniformity (of lighting and camera CCD response) and used to predict CIE *XYZ* values using either the neural network or the polynomial. The test samples were used to assess the characterization performance for the various models that were used.

For the models based upon a neural network, multilayer perceptrons (MLPs) were used that always had three input units and three output units, but the number of hidden units was varied (the implementation of a neural network for printer characterization using MATLAB's neural network toolbox is described in Chapter 9). The networks were trained using the Levenberg–Marquardt optimization procedure.

Various polynomial transforms were attempted as detailed in Table 8.2. These polynomials always attempted to map camera *RGB* values to CIE tristimulus values. A $192 \times m$ matrix was constructed from the training set where each row contained the *m RGB* terms (see Table 8.2) for one of the samples. A linear system is then assumed where the $192 \times m$ matrix is multiplied by an $m \times 3$ matrix of coefficients to yield the 192×3 matrix of tristimulus values. The values of the coefficients were determined using the training set by multiplying the pseudoinverse of the $192 \times m$ matrix of augmented *RGB* values by the 192×3 matrix of tristimulus values. Once the coefficients are determined it is trivial to compute the tristimulus values of the samples in the training set and the samples in the test set.

Figure 8.3 shows the median CIELAB colour differences of the $m \times 3$ polynomials for various values of *m* (see Table 8.2), whereas Figure 8.4 show the training and testing error for the neural networks with *n* hidden layers.

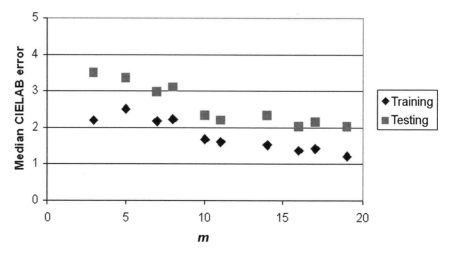

Figure 8.3   Effect of number of terms (m) in the polynomial model on training and testing performance (median colour difference)

It is evident that the performance of the best neural network and polynomial models produces a test error of about 2 CIELAB units. It is not surprising that the two systems should provide equivalent performance. Training the neural networks can be quite time consuming, however, and there are many parameters to determine, such as the number of hidden units, the transfer functions for the units in the network, the learning rule, the parameters of the learning rule, and so on and so forth. Furthermore, each time the network is trained from different random initial weights a different transform is achieved. The results shown in

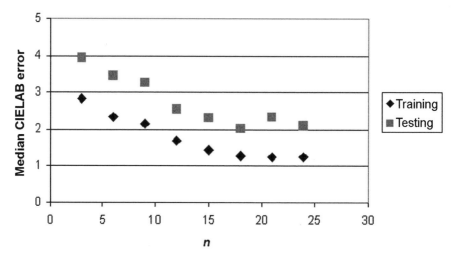

Figure 8.4   Effect of number of hidden units (n) in the neural-network model on training and testing performance (median colour difference)

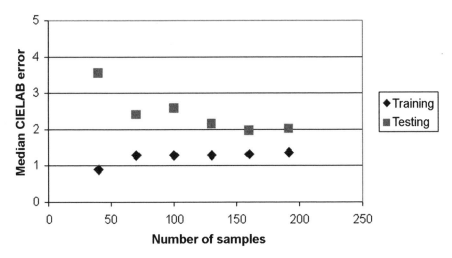

Figure 8.5 Effect of number of training samples on training and testing performance (median colour difference) for the 16 × 3 polynomial model

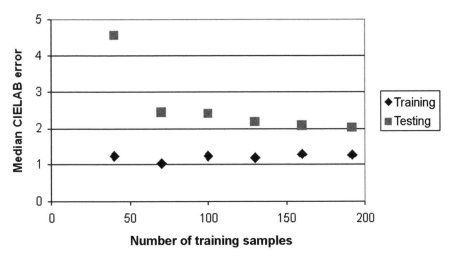

Figure 8.6 Effect of number of training samples on training and testing performance (median colour difference) for the neural network with 18 hidden units.

Figure 8.4 are in fact the average results from training each network five times. By contrast, it is relatively straightforward to develop the polynomial systems and relatively fewer decisions regarding parameters need to be made. There seems little reason therefore to use neural networks for device characterization.

Figures 8.5 and 8.6 show the effect of reducing the size of the training set for the best neural and polynomial models. The 192 samples of the full training set were progressively subsampled and new networks and polynomial systems were derived. The performance on the training set is then the median colour difference using the reduced training on which the system was trained and the performance of the test set was as before. Both systems show a surprising degree of robustness with test performance, only degrading substantially for training set sizes less than 50. Note that for the polynomial system in particular, that the training error reduces for small training set sizes. If a proper independent test set was not used to evaluate the performance of the modes, then the use of a small set of training samples could lead to an optimistic view of the performance of the characterization models.

9

Characterization of Printers

9.1 Introduction

Some general comments regarding characterization can be found in Chapter 7, Section 7.1. Physical models tend to play a more important role in the characterization of printers than they do with other imaging devices. One reason for this is that the relationship between printer inputs and CIE tristimulus values is usually extremely non-linear. In addition, however, there is a great deal of theory that has been developed to predict the colour of printing inks from colorant concentration values in a wider context. The Kubelka–Munk theory, for example, has been used for more than half a century to predict spectral reflectance from colorant concentration values. Artificial neural networks have also been used quite widely to find mappings between vectors of colorant concentration values and spectral reflectance values. Numerous technologies are used in printers and this is another reason why different and specific models are used to characterize the devices. Most printers use three or four primaries: cyan, magenta, yellow and black. Note that the primaries of a subtractive colour-mixing process are quite different from those (typically red, green and blue) for an additive colour mixing process. For both additive and subtractive devices the primaries are normally selected to enable the greatest gamut of colours to be reproduced. In a subtractive process, the intensities of the red, green and blue light in the print are indirectly controlled by the amount of the cyan, magenta, and yellow ink deposited, respectively. Some printers – typically dye-sublimation printers – operate by depositing a layer of ink where the thickness of the ink is varied to control the colour of the print. Other printers, however, such as most laser printers, use a half-tone process. For half-tone printers a fixed thickness of ink is deposited in a pattern of dots, and tonal and colour variation is achieved by varying either the size or the frequency of the dots. It is not unreasonable therefore that different physical models are used for different printers depending upon the technology that the printer uses. Nevertheless, the aim of printer

Computational Colour Science Using MATLAB. By Stephen Westland and Caterina Ripamonti.
© 2004 John Wiley & Sons, Ltd: ISBN 0 470 84562 7

characterization is the same as in camera or monitor characterization. Device coordinates (cyan, magenta, yellow and black) are converted into device-independent CIE XYZ values. In this chapter the use of physical models for the characterization of printers is described. There is particular emphasis on the Kubelka–Munk and Neugebauer models for device characterization of half-tone printers. Finally, in Section 9.5 two examples of printer characterization are detailed; one for a half-tone printer and one for a continuous-tone printer.

9.2 Physical models

Characterization of input and display devices is predominantly achieved through linear and non-linear transforms. However, although these techniques are also often used for the characterization of printers, physical models are also important for these devices. Physical printer models can be categorized into two types (Green, 2002b): (i) those that aim to predict the relationship between reflectance and dot area or colorant strength; and (ii) those that predict the colour of different colorant combinations, in terms of either colorimetry or spectral reflectance. It may be useful to consider these two models as processes of colorant and colour prediction, respectively, and to recognize that they are inversely related. Thus, many models can be used to predict reflectance or tristimulus values from colorant information but can then be inverted to predict colorant information.

Many printing systems print solid-colour ink in a dot pattern. Such half-tone systems provide tonal variation by varying either the size of the dots or their frequency. The measured reflectance of a half-tone system may be predicted by spatially averaging the colours of the dots and the substrate on which the dots are printed. A weighted average for each pixel in the image usually is computed based upon the proportional areas of the dots and the substrate. Models such as Neugebauer and Murray–Davies are used for this purpose and such models can also take into account mechanical and optical dot gain. Mechanical dot gain is the phenomenon where the printed dot is physically larger than it should be because of ink spreading during the printing process. Optical dot gain is where there is an apparent gain in the size of the dot caused by scattering the substrate. Substrate scattering is responsible for light being absorbed by the ink dot even when it strikes the substrate directly on an unprinted area. When more than one colour is printed, the second colour can overprint the first. The Neugebauer model must include the colour of the substrate, the primary colours and the overprint colours. For a typical printing system the number of possible overprint colours usually is quite small and therefore it may not be unreasonable to measure them directly. In certain circumstances, however, it may be necessary to predict the overprint colour and the Kubelka–Munk theory may be used for this purpose (Bala, 2003).

The Kubelka–Munk theory characterizes each colorant using the absorption K and scattering S coefficients at each wavelength. The theory can be difficult to apply since specific calibration samples are required to allow estimation of the K and S values (Nobbs, 1985).

9.3 Neural networks

The field of artificial neural networks (ANNs) defines a set of computational methods that were inspired from studies of how humans process information to solve problems. There are many different types of ANNs and the reader is recommended to study the extensive literature that is now available to explain the principles and algorithms of neural computing (e.g. Rumelhart and McClelland, 1986; Kohonen, 1988; Aleksander and Morton, 1991; Haykin, 1994). This chapter includes only a cursory analysis of just one class of neural network known as a multilayer perceptron (MLP). Despite the large variety of network structures that have been developed, the majority of practical applications of neural computing are in fact based upon MLPs.

An MLP consists of layers of processing units known as neurones, or simply units. Each unit receives input and performs some function upon this input to produce an output. The function between input and output for any unit is known as the activation function, or the transfer function, and normally is non-linear. A typical non-linear transfer function is the sigmoid (S-shaped) function, but linear transfer functions sometimes are used for the units in the output layer. The input for each unit is the weighted sum of the outputs from all the units in the previous layer. The units in the first layer (known as the input layer) receive their input from an input vector and those in the last layer (known as the output layer) generate an output vector. Each unit in the hidden and output units also receives weighted input from a bias unit whose output is fixed at unity. The network as a whole can be thought of as a universal function approximater that attempts to find a mapping between input vectors and output vectors (Figure 9.1). Such networks are interesting because, in principle, they can perform any valid mapping to any arbitrary degree of accuracy. A valid mapping is one that is computable.

The number of units in the input and output layers is determined from the nature of the problem being solved. If, for example, the network is being used to perform a mapping between a four-dimensional vector and a one-dimensional vector ($f : \Re^4 \to \Re^1$), then the number of units in the input and output layers would be four and one, respectively. However, the number of hidden layers and the number of units in each hidden layer must be determined empirically. For ANNs it is important to distinguish between the training mode and the testing mode. During the training mode, examples of an input–output pair are presented to the network, the error between the desired output and the actual output (using

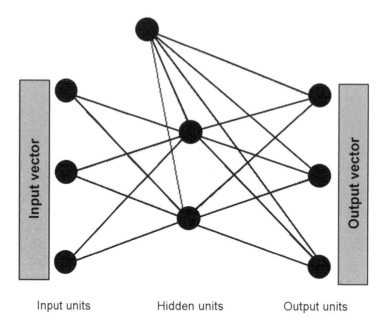

Figure 9.1 Schematic diagram for an MLP. The solid lines represent weighted connections between the processing units (●)

the current set of weights) is computed, and the values of the weights are modified to reduce this error. This process is repeated for each input–output pair in the training set and the presentation of the whole training set in this way is known as a training epoch. Training may require thousands or even hundreds of thousands of epochs and typically the training procedure is very computationally intensive. However, at the end of the training period the values of the weights are fixed. During the testing mode, input vectors are presented to the network and output vectors are computed. The performance of the network in testing mode using the data from the training set is known as the training error. A common problem with MLPs is that they are prone to over-fitting the training data. As the number of hidden layers or units in the network increases, the training error should decrease. In the limit a sufficiently complex MLP can produce a training error of zero; such a network, however, may exhibit poor generalization performance. Generalization is the ability of the network to perform using data that was not used during the training period. A second data set, known as a testing data set, is therefore used to determine the testing error. Of course, the training and testing data sets should be drawn from the same population so that they both represent, in a statistical sense, the problem being addressed by the network.

9.4   Characterization of half-tone printers

9.4.1   Correction for non-linearity

If we consider a single ink printed on a substrate in a half-tone pattern and denote the reflectance of the unprinted substrate by P_w and the reflectance of the solid ink by P_s, then the Murray–Davies relationship (Yule, 1967) predicts the measured reflectance P of the print. The value of P is related to the sum of the reflectances of the two components weighted by their fractional area coverage,

$$P = AP_s + (1 - A)P_w, \tag{9.1}$$

where A is the proportional area of the paper that is covered by ink.

Equation (9.1) can be inverted to predict the proportional dot area,

$$A = (P_w - P)/(P_w - P_s). \tag{9.2}$$

The simple Murray–Davies equation does not take dot gain into account. Yule and Nielsen (1951) proposed a correction to the Murray–Davies equation,

$$P = [A(P_s)^{1/2} + (1 - A)(P_w)^{1/2}]^2. \tag{9.3}$$

Equation (9.3) results in a non-linear relationship between the area coverage A and the resulting reflectance P. The generalized Yule–Nielsen equation allows an exponent n so that

$$P = [A(P_s)^{1/n} + (1 - A)(P_w)^{1/n}]^n, \tag{9.4}$$

where n usually is given a value between 1.0 (for a glossy substrate) and 2.0 (for a matt substrate). This non-linear relationship is required to account for the phenomenon of optical dot gain. Optical dot gain is the phenomenon that half-tone prints usually appear darker than expected [based on Equation (9.1)] because some light that strikes the unprinted substrate is absorbed by the ink dots. This occurs because of light scattering in the substrate (Figure 9.2).

In addition to optical dot gain it is also necessary to consider mechanical dot gain, which is the phenomenon where the printed dots usually are physically larger than their target sizes because of flow of the wet ink when it is applied to the substrate. The effect of mechanical dot gain is that a non-linear relationship exists between the digital input count d and the dot coverage A. For given values of d and P, the optimum area coverage A may be computed using

$$A = \sum(P_s(\lambda)^{1/n} - P(\lambda)^{1/n})(P_s(\lambda)^{1/n} - P_w(\lambda)^{1/n})/\sum(P_s(\lambda)^{1/n} - P_w(\lambda)^{1/n}), \tag{9.5}$$

where it is assumed that there are no inter-colorant interactions (Bala, 2003). Thus, P_j is measured for a number of levels d_j and then Equation (9.5) is used to determine A_j. This procedure yields pairs of $[d_j \ A_j]$ from which a continuous function can be derived that maps the digital count d to dot area coverage A.

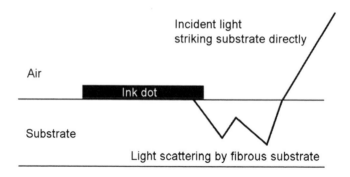

Figure 9.2 Light that strikes the unprinted area of the substrate may be absorbed by the ink because of scattering of light within the substrate

Some alternative methods for determining dot areas that minimize the error in CIELAB colour-difference units are also available (Bala, 1999).

9.4.2 Device-independent representation

For half-tone printers, device-independent representation is normally obtained by finding a mapping between the proportional dot coverages for the inks and the spectral reflectance of the print, from which it is then trivial to compute XYZ values. Alternative methods may use neural networks to find a mapping either from dot coverages to reflectance or even directly to CIE XYZ values. The most common method for predicting reflectance involves the Neugebauer model which takes into account the various overlapping binary mixtures. For example, if cyan, magenta and yellow inks are considered, then the resulting reflectance will be a function of the reflectances of the unprinted substrate P_w, the three solid colours (P_c, P_m, P_y) and the four overlap colour combinations of cyan + magenta (blue P_b), cyan + yellow (green P_g), yellow + magenta (red P_r) and black P_k. If the fractional areas of these eight areas are represented by A_c, A_m, etc. then we can write

$$P = A_w P_w + A_c P_c + A_m P_m + A_y P_y + A_b P_b + A_g P_g + A_y P_y + A_k P_k. \quad (9.6)$$

It is evident that the Neugebauer model is a straightforward extension of the Murray–Davies equation [Equation (9.1)] which assumes that the reflectance of a spatial area is the additive combination of the reflectances of the primary colours and their overlapping areas. In the original Neugebauer equations the approach was used to predict the broadband reflectance in the short-, medium- and long-wavelength portions of the spectrum and, indeed, modern versions of Neugebauer sometimes operate using XYZ tristimulus values. However the n-modified spectral Neugebauer approach [illustrated for a CMY system by

Equation (9.6)] has been shown to be most accurate (Bala, 2003). In the n-modified Neugebauer model all the reflectances are raised to the power $1/n$ as in Equation (9.5).

In order to implement the Neugebauer approach the digital counts are first converted into the dot coverage areas using a tone-reproduction curve, as described in the previous section. A method to compute the actual areas of the primary and secondary colours is then required. For the three-colour example, the proportional areas of the eight colour regions can be computed using Demichel's equation (Green, 2002c),

$$
\begin{aligned}
A_w &= (1 - c)(1 - m)(1 - c), \\
A_c &= c(1 - m)(1 - y), \\
A_m &= m(1 - c)(1 - y), \\
A_y &= y(1 - c)(1 - m), \\
A_b &= cm(1 - y), \\
A_g &= cy(1 - m), \\
A_r &= my(1 - c), \\
A_k &= cmy,
\end{aligned}
\tag{9.7}
$$

where c, m and y are the proportional dot areas of the three primary colours obtained from the tone-reproduction curves. Demichel's equation has been shown (Viggiano, 1990) to work reasonably well for rotated half-tone screen configurations where the screens for cyan, magenta and yellow are placed at different angles that are carefully selected to avoid moiré artifacts.

It is important to note, however, that Equations (9.7) make certain assumptions concerning the amount of overlap between the primary colours. If we consider the case where $c = 0.4$, $y = 0.4$ and $m = 0$, then Demichels's equation will predict $A_w = 0.36$, $A_c = 0.24$, $A_y = 0.24$ and $A_g = 0.16$. However, it would be possible for the cyan and magenta dots to be printed without overlap ($A_w = 0.20$, $A_c = 0.40$, $A_y = 0.40$ and $A_g = 0.00$), with total overlap ($A_w = 0.60$, $A_c = 0.00$, $A_y = 0.00$ and $A_g = 0.40$) or with any intermediate amount of overlap. The primaries normally are printed at different screen angles and the relationship between these two angles is one of several factors that could affect the degree of overlap. The dot-on-dot half-tone configuration, for example, places the primary dots at the same screen angle and phase so that they maximally overlap. In practice it has been shown that a weighted combination of the Demichel model and the dot-on-dot model can give good performance (Bala, 2003).

9.4.3 The Kubelka–Munk model

The Neugebauer models assume that the reflectance (for spectral Neugebauer approaches) or tristimulus values (for tristimulus Neugebauer approaches) are known for the over-printed area of the secondary colours. So, for example, to implement where yellow dots are over-printed with cyan dots we need to know the colour of the over-print area where cyan ink falls on yellow ink.

When inks are printed on top of each other or are mixed together and then printed, subtractive colour mixing takes place and additivity of reflectance values is not valid. For subtractive mixing the densities of the inks are approximately additive, where the density D is related to the reflectance P,

$$D = -\log_{10} P. \tag{9.8}$$

So, for example, if two inks have reflectance 0.4 and 0.8 at a certain wavelength and they are mixed together in equal proportions, then the mean of the density contributions will be $0.0969/2 + 0.3979/2 = 0.2474$ corresponding to a reflectance of 0.566 (this compares with a value of 0.600 if the reflectances are directly averaged). Accurate prediction for subtractive mixing often requires application of the Kubelka–Munk theory of radiation transfer that characterizes each ink or colorant in terms of its absorption and scattering properties.

The Kubelka–Munk theory (Nobbs, 1985, 1997; McDonald, 1997b) has been used to predict the reflectance of inks, plastics, paints, textiles and other materials. The theory characterizes each colorant using the absorption K and scattering S coefficients that are functions of wavelength and relates these coefficients to the body reflectance of a sample. The body reflectance is the reflectance of a surface if the interactions of light at the air/medium interface are discounted. The body reflectance R is related to the measured reflectance P by the following equation,

$$R(\lambda) = [P(\lambda) - r_e]/[(1 - r_e)(1 - r_i) + r_i(P(\lambda) - r_e)] \tag{9.9}$$

for the case where P is measured with a spectrophotometer with the specular component included. The variables r_e and r_i are the external and internal reflectance coefficients of the boundary. The inverse of Equation (9.9) is given by Equation (9.10),

$$P(\lambda) = r_e + [(1 - r_e)(1 - r_i)R(\lambda)]/[1 - r_e R(\lambda)]. \tag{9.10}$$

For an opaque sample the body reflectance is related to the K and S coefficients by Equation (9.11),

$$K/S = (1 - R)^2/2R, \tag{9.11}$$

and the inverse relationship is given by

$$R = 1 + K/S - [(1 + K/S)^2 - 1]^{1/2}. \tag{9.12}$$

Thus, for opaque samples only the ratio of K to S is required at each wavelength in order to predict the reflectance R. In the case of dyed textiles the dyes themselves do not scatter light and the only scattering comes from the textile fibres to which the dyes are applied. The application of the Kubelka–Munk theory to opaque dyed textiles is therefore often referred to as the one-constant version of the theory. For many pigmented surface coatings, such as paints, the pigments both absorb and scatter and the two-constant theory is required. However, Equation (9.12) can still be used to predict the reflectance of the surface coating if it is applied at a thickness that achieves opacity. For translucent printing inks, however, the reflectance of the paper upon which the ink is printed makes a contribution to the reflectance of the system and therefore Equation (9.12) must be replaced by

$$R = [(R_g - R_\infty)/R_\infty - (R_\infty R_g - 1)\exp\{(1/R_\infty - R_\infty)Sx\}]/$$
$$[(R_g - R_\infty) - (R_g - 1/R_\infty)\exp\{(1/R_\infty - R_\infty)Sx\}], \quad (9.13)$$

where R_g is the reflectance of the substrate and R_∞ is the reflectance [as defined by Equation (9.12)] of an opaque layer of the pigmented layer. The scattering coefficient S is defined for a unit thickness of the layer and x is the thickness of the layer. According to the theory the values of K and S should be linearly related to the pigment volume concentration in the layer and to the thickness of the layer. However, in practice severe departures from linearity can occur (Nobbs, 1997). In order to predict the reflectance for a mixture of colorants or inks the K and S contributions are determined for each component and then assumed to be additive in order to allow the computation of K and S for the layer and thus, via Equation (9.13), the reflectance R. In all cases, once the body reflectance is known Equation (9.11) can be used to yield a prediction of the reflectance P.

The Kubelka–Munk theory is routinely used for the prediction of reflectance for systems of printing inks (for example, in lithography) and forms the basis of computer match-prediction systems. However, one of the difficulties in applying the theory to the characterization of printers is in determining the values of K and S for the individual inks. One method to determine K and S is to print each colorant over two different substrates or papers (for example, a white and a black) and then to use Equation (9.13) to set up a system of two simultaneous equations with two variables (K and S). For many printing systems it is difficult to obtain these samples, especially since it is required that the surface properties (roughness, etc.) of the two substrates must be identical. An alternative approach is to treat the Kubelka–Munk coefficients as free parameters and to derive their values based on an optimization routine and a set of samples of known reflectance (Bala, 2003).

The Kubelka–Munk model could be used to predict the overlap areas in a half-tone printing process and Neugebauer-type models could then be used to predict the reflectance of a given area. The traditional Kubelka–Munk theory assumes that the printed layer is homogeneous, however, whereas for half-tone printing one ink is printed on top of another to generate a more layered

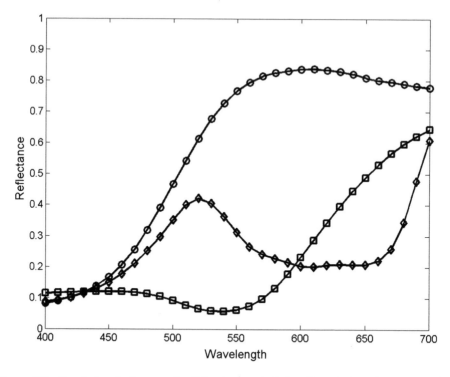

Figure 9.3 Spectral reflectances of solid colour red (□), yellow (○) and green (◇) inks printed by a half-tone process over white

structure. Nevertheless, applications of modified Kubelka–Munk models to the characterization of printers have been reported in the literature (Kang, 1994; Emmel and Hersch, 2000).

9.5 Implementations and examples

This section contains two examples of printer characterization; one for a half-tone printer and one for a continuous-tone printer.

9.5.1 Half-tone printer

Figure 9.3 shows the reflectance spectra for three inks printed at full coverage (solid colour) over white using a half-tone process. In order to characterize the tone-reproduction curves of the three inks each was printed and measured at target digital coverage proportional areas between 0 and 1 at intervals of 0.1. The function *gettrc* uses Equation (9.5) to compute the actual area coverage for each target area coverage given the value of n.

Box 25: *gettrc.m*

```
function [p] = gettrc(dig,R,W,Solid,n,graphs);

% function [p] = gettrc(dig,R,W,Solid,n,graphs)
% gettrc function to compute the trc for an ink
% function [p] = gettrc(dig,R,W,Solid,n,graphs);
% dig is an 1 by n matrix of target area coverages
% R is an n by m matrix of measured reflectance values
% W is a 1 by m matrix of reflectance for the white substrate
% Solid is a 1 by m matrix of reflectance for the solid ink
% n is a free parameter > 0
% graphs = 'on' for graphical display
% p is a matrix containing the coefficients of a polynomial
% to relate target coverage to actual coverage

if nargin < 6
  plotgraphs = 0;
else
  plotgraphs = strcmp('on',graphs);
end

graphs = 1;
num = length(dig);

R = R.^(1/n);
W = W.^(1/n);
Solid = Solid.^(1/n);

for i = 1:num
  c(i) = sum((Solid - R(i,:)).*(Solid - W))/sum((Solid ...
- W).*(Solid - W));
end

c = 1-c;

[p,s] = polyfit(dig,c,3);

if (plotgraphs)
  figure
  plot(dig,c,'k*')
```

```
x = linspace(0,1,101);
y = polyval(p,x);
hold on
plot(x,y,'k-')
end
```

The built-in MATLAB function *polyfit* is then used to fit a third-order polynomial between the actual and target coverage areas and the coefficients of this fit are returned in the matrix **p**. The full syntax for *gettrc* is

```
function [p] = gettrc(dig,R,w,solid,n,graphs)
```

where **dig** is a $1 \times r$ matrix of target area coverages, **R** is an $r \times m$ matrix of measured reflectance values, **w** is a $1 \times m$ matrix of reflectance for the white substrate, **solid** is a $1 \times m$ matrix of reflectance for the solid ink, **n** is a 1×1 matrix containing the free parameter [see Equation (9.5)] to determine the non-linearity, r is the number of target coverage areas, m is the number of wavelengths (usually 31) at which the reflectance data are measured and **graphs** = 'on' causes a plot of the actual versus target areas to be generated. Once the characteristics of the tone-reproduction curve have been established using *gettrc* any target coverage can be converted to the actual coverage using the *polyval* command:

```
actual = polyval(p,target)
```

where **target** and **actual** are 1×1 matrices that hold the target and actual area coverage values and p is the output of the *gettrc* function. Figures 9.4 and 9.5 show the result of the *gettrc* function for $n = 1$ and $n = 20$, respectively, for each of the three inks.

The optimum value of n may be found to give the lowest prediction error for the reflectances of the inks printed on their own at various coverage areas and in mixture with other inks. For this example application, a value of $n = 20$ was found to be optimum. The tone reproduction curves allow the target areas to be converted to actual areas and these may be used with the n-modified Neugebauer equations to predict the spectral reflectance for combinations of inks. For a single ink, and measurements at 31 wavelengths between 400 and 700 nm, the colour difference between the predicted reflectance and the measured reflectance is predicted using the following code:

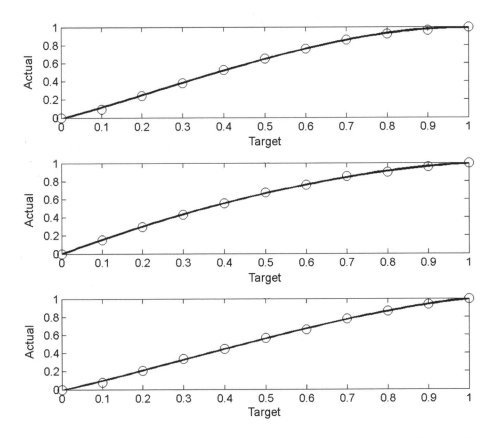

Figure 9.4 Target and actual coverage areas with $n = 1$ for red (upper panel), yellow (middle panel) and green (lower panel) inks. Solid lines show the polynomial fits

```
c = polyval(p,dig);
for w = 1:31
    pred(w) = (c*(solid(w))^(1/n)+(1-c)*(W(w)^(1/n)))^n;
end
xyzm = r2xyz(meas, 400, 700, 'd65_64');
xyzp = r2xyz(pred, 400, 700, 'd65_64');
labm = xyz2lab(xyzm, 'd65_64');
labp = xyz2lab(xyzp, 'd65_64');
thisDE = cmcde(labm, labp)
```

where **solid** and **w** are 1×31 matrices containing the reflectances of the ink printed at full coverage and the white substrate itself, **dig** is a 1×1 matrix containing the target proportional area coverage of the ink for this print, **p** contains the tone-reproduction curve for this ink (see the *gettrc* function) and **n** is a 1×1 matrix containing the degree of linearity.

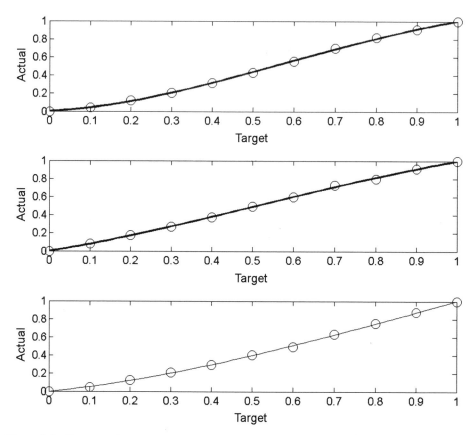

Figure 9.5 Target and actual coverage areas with $n = 20$ for red (upper panel), yellow (middle panel) and green (lower panel) inks. Solid lines show the polynomial fits

For a binary mixture, the following code would be used:

```
c1 + polyval(p1, dig1);
c2 polyval(p2, dig2);

% get the areas using Dimechel
A1 = c1*(1-c2);
A2 = c2*(1-c1);
Aw = (1-c1)*(1-c2);
Ao = c1*c2;

for w = 1:31
  pred(w) = (Ab*(solid1(w))^(1/n) + ...
    At*(solid2(w))^(1/n) + ...
    Ao*(overlap(w))^(1/n) + Aw*(W(w))^(1/n))^n;
```

```
end
xyzm = r2xyz(meas, 400, 700, 'd65_64');
xyzp = r2xyz(pred, 400, 700, 'd65_64');
labm = xyz2lab(xyzm, 'd65_64');
labp = xyz2lab(xyzp, 'd65_64');
thisDE = cmcde(labm, labp)
```

where **c1** and **c2** are the target area coverages for the inks printed first and second, respectively, **solid1** and **solid2** are the solid ink colours for the two inks, **p1** and **p2** are the respective tone-reproduction curves and **overlap** is a 1×31 matrix containing the reflectance of the overlap region. The entries of **overlap** may be measured from a print of the solid colour of one ink over another or it could be predicted using a Kubelka–Munk (or some other) model.

9.5.2 Continuous-tone printer

An example of the characterization of a continuous-tone printer is provided from Sueeprasan (2003) who used a third-order masking model to characterize a Kodak Color Proofer 9000A printer that is based upon dye-sublimation technology. The data collected by Sueeprasan is used here to directly compare characterization models based upon neural networks and polynomial trans-forms. Two sets of data were obtained: a set of training data contained 729 colours and a set of test data contained 144 colours. For both sets the *RGB* inputs to the printer driver were available and the CIE *XYZ* values for illuminant D65 were measured (Figures 9.6 and 9.7).

In this example characterization, mappings are developed between the printer *RGB* values and the CIE *XYZ* values. For many practical situations a mapping that predicts *RGB* values based upon *XYZ* values would be more useful. To test such a mapping requires that the *RGB* values predicted by the mapping are physically reproduced using the imaging device and the CIE vales measured and compared with the target values. This testing procedure is necessary in many cases but is time consuming. For this example, the advantage of the $RGB \rightarrow XYZ$ mapping is that a single test set can be created and used to evaluate various characterization methods. The *XYZ* values of the test set are known. Evaluation of the methods using the test set is achieved simply by comparing the *XYZ* values from the model with the actual *XYZ* values that were measured for the test set.

A characterization was first developed using a neural network and the training samples. The input to the network was a 3×729 array of *RGB* values. The original range of *RGB* values was [0, 255] but they were scaled to be in the range [0, 1] before presenting them to the network. The output to the network was a 3×729 array of $L^*a^*b^*$ values and these were also scaled to be approximately in the range [0.1, 0.9]. It is important to take into account the range of the output

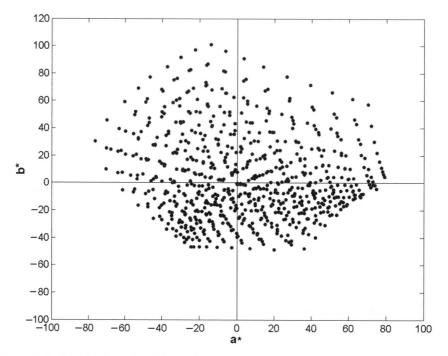

Figure 9.6 Distribution of training colours in CIELAB space for dye-sublimation printer characterization example

vectors and the nature of the transfer or activation function on the units of the output layer since some functions are only able to output data in a certain range. The sigmoid activation function, for example, which was used in this study can only output values in the range [0, 1] and the extreme values of this range are only achieved with input values to the function that are infinitely large. For use with the sigmoid activation function it is quite common to scale the output vectors to a range [0.1, 0.9]. An MLP network was used with a single hidden layer. The number of units in the input and output layers was three and, initially, six units were used in the hidden layer. The network was created using the MATLAB command

```
net =newff([0 1; 0 1; 0 1], [6, 3], {'logsig' 'logsig'});
```

which creates a feed-forward network (or MLP). The first argument to the function specifies that there are three input units and declares the range of values that are expected (this allows appropriate scaling to be automatically carried out if the data do not span an appropriate range). The second argument to the function specifies that there is one hidden unit and declares the number of units in the hidden and output layers. The final argument declares the use of the sigmoid activation function for the hidden and output layers.

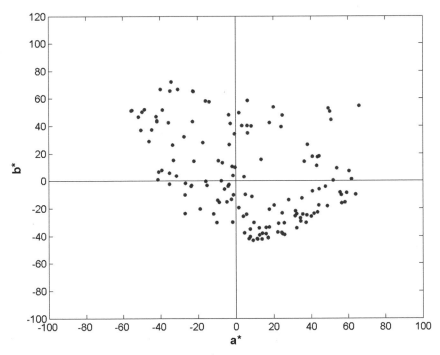

Figure 9.7 Distribution of test colours in CIELAB space for dye-sublimation printer characterization example

The following two MATLAB commands set the weights of the network to initial values and specify the number of epochs for training,

```
net = init(net);
```

```
net.trainParam.epochs = 1000;
```

The network can now be trained using the single command

```
net = train(net, input, output);
```

where the **input** and **output** matrices are the 3×729 arrays of *RGB* and *L*a*b** values, respectively. During the training process MATLAB generates a graph showing how the error between the actual and predicted output matrices changes with the number of epochs that have elapsed. An example of that graph is illustrated by Figure 9.8 for one particular training run and it is evident that most of the training took place in the first few hundred epochs. The default training algorithm is based upon Levenberg–Marquardt optimization, which is an extremely efficient method for training an MLP.

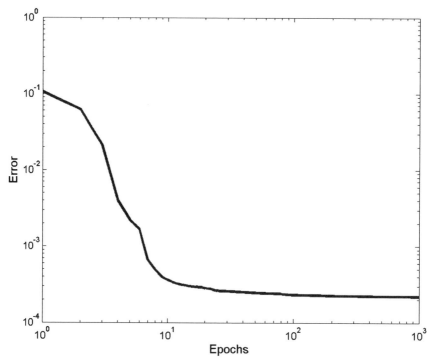

Figure 9.8   Typical learning behaviour of a neural network trained to map from *RGB* to *L*a*b**

After training the network performance was tested using the MATLAB command

```
poutput = sim(net,input);
```

which generates the predicted output matrix poutput for the matrix **input** given the state of the network **net**. The target and predicted output values were re-scaled to the original values of the CIELAB space and the colour difference was computed for each of the 729 samples. The median CIELAB colour difference was 3.47 (maximum 12.89). This error is referred to as the training or memorization error. The testing or generalization error was computed by using the *sim* command with the training input matrix containing the 144 samples and the median colour difference was found to be 3.16 (maximum 11.87). There is no reason why, as in this case, the testing error may not be less than the training error. However, the testing error should always be used as the indication of the network's ability to predict output for input vectors that were not used during the training process.

The number of units in the hidden layer was varied and the above process repeated. Table 9.1 lists the performances that were obtained. The performance of the neural network was compared with that of a third-order masking or

Table 9.1 CIELAB errors for MLP to map $RGB \rightarrow L^*a^*b^*$ compared with a third-order polynomial model

	Memorization			Generalization		
Layers	Minimum	Median	Maximum	Minimum	Median	Maximum
4	0.34	4.37	16.59	0.41	3.90	12.14
6	0.28	3.84	12.89	0.41	3.16	11.87
8	0.31	2.92	9.43	0.44	3.84	11.36
10	0.21	2.81	9.96	0.37	4.23	11.04
12	0.21	2.29	9.32	0.77	4.29	10.28
Polynomial						
20×3	0.37	3.99	9.99	0.52	4.01	10.59

polynomial model. The model used by Sueeprasan (2003) was a third-order masking model that predicted colorimetric densities from printer RGB values. The colorimetric densities were computed by the terms $\log(X/X_n)$, $\log(Y/Y_n)$ and $\log(Z/Z_n)$, where the subscript n referred to the white point (as can be seen from Table 4.2 the white point for illuminant D65 and the 1931 observer is $X_n = 95.047$, $Y_n = 100.00$ and $Z_n = 108.883$). The mapping was accomplished by the linear system

$$\mathbf{P} = \mathbf{AC}, \tag{9.14}$$

where $\mathbf{P}$ is a 729×3 matrix of $1 - R/255$, $1 - G/255$ and $1 - B/255$ terms for each of the 729 training samples, $\mathbf{A}$ is a 729×20 matrix of augmented colorimetic densities and $\mathbf{C}$ is a 20×3 matrix of coefficients that defines the mapping. Each row of the augmented matrix contains the following terms: R, G, B, RG, RB, GB, R^2, G^2, B^2, R^2G, R^2B, G^2R, G^2B, B^2R, B^2G, R^3, G^3, B^3, RGB and 1. The coefficients $\mathbf{C}$ were determined using

$$\mathbf{C} = \mathbf{A}^+\mathbf{P}, \tag{9.15}$$

which minimizes the least-squared error between the target and predicted colorimetric densities.

The following code illustrates how the third-order model was implemented and tested,

```
clear
load train.mat
% trainrgb is a 729 by 3 matrix of RGB values
% trainxyz is a 729 by 3 matrix of XYZ values
load test.mat
% testrgb is a 144 by 3 matrix of RGB values
```

```
% testxyz is a 144 by 3 matrix of XYZ values

white = [95.047 100.000 108.883];

trainrgb = 1 - trainrgb/255;
testrgb = 1 - testrgb/255;

traintarget = zeros(729,3);
for i = 1:729
  traintarget(i,:) = log(trainxyz(i,:)./white);
end

% the construction of the augmented matrix is shown in
% full below for clarity
trainmat = zeros(729,20);
for i = 1:729
  trainmat(i,1) trainrgb(i,1);
  trainmat(i,2) = trainrgb(i,2);
  trainmat(i,3) = trainrgb(i,3);
  trainmat(i,4) = trainrgb(i,1)*trainrgb(i,2);
  trainmat(i,5) = trainrgb(i,1)*trainrgb(i,3);
  trainmat(i,6) = trainrgb(i,2)*trainrgb(i,3);
  trainmat(i,7) = trainrgb(i,1)*trainrgb(i,1);
  trainmat(i,8) = trainrgb(i,2)*trainrgb(i,2);
  trainmat(i,9) = trainrgb(i,3)*trainrgb(i,3);
  trainmat(i,10) = trainrgb(i,1)*trainrgb(i,1*...
    trainrgb(i,2);
  trainmat(i,11) trainrgb(i,1)*trainrgb(i,1)*...
    trainrgb(i,3);
  trainmat(i,12) = trainrgb(i,2)*trainrgb(i,2)*...
    trainrgb(i,1);
  trainmat(i,13) = trainrgb(i,2)*trainrgb(i,2)*...
    trainrgb(i,3);
  trainmat(i,14) = trainrgb(i,3)*trainrgb(i,3)*...
    trainrgb(i,1);
  trainmat(i,15) = trainrgb(i,3)*trainrgb(i,3)*...
    trainrgb(i,2);
  trainmat(i,16) = trainrgb(i,1)*trainrgb(i,1)*...
    trainrgb(i,1);
  trainmat(i,17) trainrgb(i,2)*trainrgb(i,2)*...
    trainrgb(i,2);
  trainmat(i,18) = trainrgb(i,3)*trainrgb(i,3)*...
    trainrgb(i,3);
```

```
    trainmat(i,19) = trainrgb(i,1)*trainrgb(i,2)*...
      trainrgb(i,3);
    trainmat(i,20) = 1;
end

% compute the coefficients using the pinv command
a = pinv(trainmat)*traintarget;

% now implement the model
ptraintarget = trainmat*a;
% convert the predicted densities back to XYZ values
for i = 1:729
  ptraintarget(i,:) = exp(ptraintarget(i,:)).*white;
end

% compute CIELAB Delta E values
de = zeros(729,1);
for i = 1:729
  lab1 = xyz2lab(trainxyz(i,:),'d65_31');
  lab2 = xyz2lab(ptraintarget(i,:),'d65_31');
  de(i,:) = cielabde(lab1, lab2);
end

% no semicoln on this line so that the results are displayed
perf = [min(de) mean(de) max(de)]
% compute the augmented matrix for the test set
testmat = zeros(729,20);
for i = 1:144
  testmat(i,1) = testrgb(i,1);
  testmat(i,2) = testrgb(i,2);
  testmat(i,3) = testrgb(i,3);
  testmat(i,4) = testrgb(i,1)*testrgb(i,2);
  testmat(i,5) = testrgb(i,1)*testrgb(i,3);
  testmat(i,6) = testrgb(i,2)*testrgb(i,3);
  testmat(i,7) = testrgb(i,1)*testrgb(i,1);
  testmat(i,8) = testrgb(i,2)*testrgb(i,2);
  testmat(i,9) = testrgb(i,3)*testrgb(i,3);
  testmat(i,10) = testrgb(i,1)*testrgb(i,1)*
    testrgb(i,2);
  testmat(i,11) = testrgb(i,1)*testrgb(i,1)*
    testrgb(i,3);
  testmat(i,12) = testrgb(i,2)*testrgb(i,2)*
    testrgb(i,1);
```

```
    testmat(i,13) = testrgb(i,2)*testrgb(i,2)*...
      testrgb(i,3);
    testmat(i,14) = testrgb(i,3)*testrgb(i,3)*...
      testrgb(i,1);
    testmat(i,15) = testrgb(i,3)*testrgb(i,3)*...
      testrgb(i,2);
    testmat(i,16) = testrgb(i,1)*testrgb(i,1)*...
      testrgb(i,1);
    testmat(i,17) = testrgb(i,2)*testrgb(i,2)*...
      testrgb(i,2);
    testmat(i,18) = testrgb(i,3)*testrgb(i,3)*...
      testrgb(i,3);
    testmat(i,19) = testrgb(i,1)*
      testrgb(i,2)*testrgb(i,3);
      testmat(i,20) = 1;
end

% implement the model for the test set
ptesttarget = testmat*a;

% convert the densities to XYZ values
for i = 1:144
  ptesttarget(i,:) = exp(ptesttarget(i,:)).*white;
end

% compute the CIELAB Delta E values
de1 = zeros(144,1);

for i = 1:144
  lab1 =; xyz2lab(testxyz(i,:),'d65_31');
  lab2 = xyz2lab(ptesttarget(i,:),'d65_31');
  de1(i,:) = cielabde(lab1, lab2);
end

% display the test results
perf = [min(de1) mean(de1) max(de1)]
```

The results of the third-order model are shown in Table 9.1, and it can be seen that the performance is rather similar to the performance of the neural network.

10

Multispectral Imaging

10.1 Introduction

The characterization of a digital colour camera so that the device-dependent *RGB* values may be transformed to device-independent coordinates such as *XYZ* values effectively converts the camera into an imaging colorimeter, and this has many practical uses. However, in imaging science there are limitations to this approach. Many applications require that some illuminant-independent measure, such as the spectral reflectance values, be determined at each pixel location in a scene and this may be achieved by using an imaging spectro-photometer. Although imaging spectrophotometers are becoming commercially available often they are expensive and there is current interest in exploring to what extent spectral values may be recovered from a standard three-channel camera system or from a camera system with relatively few channels. The term multispectral imaging sometimes is used to define this field of research. This definition is confusing, however, since even the normal *RGB* image representation may be described as being multispectral in some sense. In this chapter, however, the term multispectral imaging will be used to define techniques and methods that may be used to recover spectral information from camera systems with a small number of channels (typically in the range 3–8). We distinguish multispectral imaging from the term hyperspectral imaging which we use to describe techniques where spectral values are measured using imaging devices with a large number of channels (typically in the range 16–40). Clearly there are situations where the distinction between multispectral and hyperspectral may become blurred. Indeed, we note that the goal of both multispectral imaging and hyperspectral imaging is the same, namely to recover a spectral image. The reader is directed to Hardeberg (2001) for an excellent review of this area. We begin this chapter with a brief review of some computational approaches to the problem of colour constancy because many of the methods of multispectral imaging were, in fact, inspired by a computational analysis of the problem of

---

Computational Colour Science Using MATLAB. By Stephen Westland and Caterina Ripamonti.
© 2004 John Wiley & Sons, Ltd: ISBN 0 470 84562 7

colour constancy. The problem of whether the visual system might be able to recover the spectral properties of objects in a scene from the cone excitations has been studied extensively and analyses of this problem are relevant for multispectral imaging. We describe some computational procedures for spectral recovery using multispectral imaging and finally describe some applications of these procedures for reflectance recovery and camera characterization.

10.2   Computational colour constancy and linear models

Colour constancy, the phenomenon by which surfaces tend to retain their approximate daylight colour appearance when viewed under a wide range of different light sources, was described in Chapter 6. It is still a mystery how the visual system is able to discount the effect of the illumination when the colour signal that reaches the eye depends just as much on the spectral power distribution of the illuminant as it does on the spectral reflectance of the surfaces in the scene (Hurlbert, 1991). One possible mechanism that could account for colour constancy is adaptation of the light receptors or cones. Such a possibility was first put forward by Von Kries and is consistent with the chromatic-adaptation transforms that were described in Chapter 6. However, adaptation is a relatively slow process and yet colour constancy seems to occur almost instantaneously as we move from one light source to another in our everyday lives. An alternative approach to adaptation was postulated by Land and McCann (1971) who suggested that the visual system may use some computational process to recover signals that are independent of the illumination in a scene. In their computational analysis, known as the Retinex theory, Land and McCann called these signals lightnesses, biological correlates of reflectance that were computed by each of the three channels in the visual system. Later, the term integrated reflectance was introduced (McCann *et al.*, 1976) to describe the illuminant-invariant signals. A number of researchers have since investigated to what extent the visual system might actually be able to recover spectral reflectances for points in a scene from the corresponding triplets of cone excitations.

There are serious limitations on what we can achieve when we set out to estimate surface reflectance from cone excitations. For example, theoretically there is an infinite number of combinations of the surface-reflectance functions P and illuminant power distributions E that could produce a given colour signal S. In addition, the visual system does not measure S directly, but rather it encodes the absorption rates of the three different cone types. It seems that if P and E were not constrained in some way, then the cone excitations would provide little useful information; fortunately there are some strong constraints on both P and E.

Suppose we have a device with three colour sensors, whose spectral responsivities are $R_k(\lambda)$, $k = 1,2,3$. The three sensor responses for a colour signal $S(\lambda)$ will be

$$r_1 = \sum_\lambda R_1(\lambda)S(\lambda),$$

$$r_2 = \sum_\lambda R_2(\lambda)S(\lambda),$$ \hfill (10.1)

$$r_3 = \sum_\lambda R_3(\lambda)S(\lambda),$$

which we can group into a single matrix equation:

$$\mathbf{r} = \mathbf{Ms},$$ \hfill (10.2)

where $\mathbf{r}$ is a 3×1 matrix containing the sensor responses, the rows of the 3×31 matrix $\mathbf{M}$ are the sensor spectral responsivities and the 31×1 matrix $\mathbf{s}$ is the colour signal (this assumes that the spectral sensitivities of the channels and the spectral power of the colour signal are represented at 31 wavelengths in the visible spectrum). The key question is: Is it possible to compute $\mathbf{s}$, given both $\mathbf{r}$ and $\mathbf{M}$? Mathematically, we can rearrange Equation (10.2) by multiplying each side of the equation by the pseudoinverse of the matrix $\mathbf{M}$ so that

$$\mathbf{s} = \mathbf{M}^+\mathbf{r}.$$ \hfill (10.3)

Unfortunately, since $\mathbf{M}$ is not a square matrix it is not trivial to compute the inverse and computational procedures must be used to estimate the inverse matrix $\mathbf{M}^+$. Estimates of $\mathbf{s}$ from Equation (10.3) are likely to be widely inaccurate. In simple terms, Equations (10.2) and (10.3) represent a set of three simultaneous equations with 31 unknowns; in mathematical terms this is an under-determined system.

To simplify the problem, imagine that the surface is viewed under an equal energy light source so that $E(\lambda) = 1$ for all λ. We can now write

$$\mathbf{r} = \mathbf{Mp},$$ \hfill (10.4)

where the 31×1 matrix $\mathbf{p}$ is the spectral reflectance. Although it may still appear that three sensor responses are insufficient to estimate $\mathbf{p}$ we know there are strong constraints on the variability of the spectral reflectances of surfaces (Maloney, 1986). [There are also known constraints (Judd et al., 1964) on the variability of natural daylight.]

The natural constraints on surfaces and lights are usefully illustrated through their representation by linear models. Typically, a set of basis functions B_j (where j is 1, ..., n) are defined such that, for example, each reflectance spectrum P_i is defined by a linear sum of the basis functions,

$$P_i = B_j a_{i,j},$$ \hfill (10.5)

where $a_{i,j}$ is the weight of the jth basis function for the ith sample. The basis functions are themselves functions of wavelength but are not constrained to be between the range [0,1] nor even to be positive at all wavelengths. The number of basis functions n usually is quite small and the weights for each reflectance spectrum define a projection of the reflectance spectrum onto the n-dimensional space of the basis functions. Such linear models of reflectance spectra and illuminant power distributions are useful because they provide an efficient method for representing and storing P and E. The linear models are also useful because they lead to simple estimation algorithms for P and E given the three sensor responses $\mathbf{r}$ ($\mathbf{r}$ could be the responses of the cones in the human visual system or the responses of a trichromatic imaging system).

We can therefore rewrite Equation (10.4) as

$$\mathbf{r} = \mathbf{MBa}, \tag{10.6}$$

where the columns of the 31×3 matrix $\mathbf{B}$ hold the first three basis functions of a linear model of reflectance spectra and the 3×1 matrix $\mathbf{a}$ holds the weights that define the particular spectrum that we are trying to recover (note that $\mathbf{p} = \mathbf{Ba}$). If we group together the term $\mathbf{MB}$ (multiplying a 3×31 matrix by a 31×3 matrix), then we can see that this is a 3×3 matrix whose entries are all known. The only unknown is $\mathbf{a}$, the weights. We can therefore rearrange Equation (10.6) to produce

$$\mathbf{a} = (\mathbf{MB})^{-1}\mathbf{r}, \tag{10.7}$$

which allows $\mathbf{a}$ to be computed by standard procedures. Once $\mathbf{a}$ has been determined the reflectance spectrum can be recovered using $\mathbf{p} = \mathbf{Ba}$. This analysis illustrates two aspects of the role of linear models. First, linear models represent a priori knowledge about the likely set of inputs. Linear models may be used to allow spectral information to be recovered from three sensor responses. Secondly, linear models work smoothly with the imaging equations. Since the imaging equations are linear, the estimation methods remain linear and simple.

Figure 10.1 shows a set of five typical reflectance spectra (Westland *et al.*, 2000) and it is clear that generally they are smooth functions of wavelength. In fact, the spectra illustrated by Figure 10.1 were measured for surfaces of natural objects (leaves, petals, etc.) but the reflectance of the surface of the output of a CMYK printer or a painted sample would most likely appear similarly smooth. This is because the smoothness originates from fundamental mechanisms by which matter interacts with light (Maloney, 1986).

An alternative way to represent the constraints of surface reflectance spectra is by their Fourier representations which are found to be band limited. Thus, if the Fourier amplitude spectrum is computed for a reflectance spectrum the energy quickly falls off with increasing spectral frequency (spectral frequency typically is expressed in units of cycles per nanometer). Above the band limit there is zero

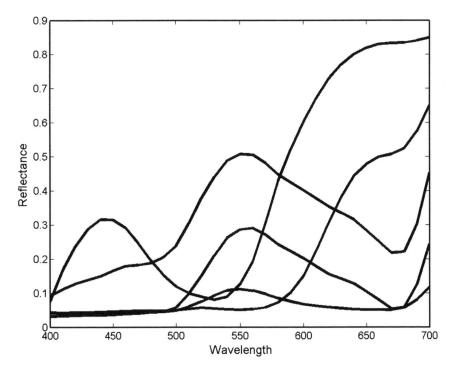

Figure 10.1 Typical reflectance spectra for five natural surfaces

energy. Estimates of the band limit for natural and man-made surfaces are in the region 0.15–0.20 cyc/nm (Maloney, 1986; Westland *et al.*, 2000).

The basis functions that describe a particular set of reflectance spectra can be obtained using a procedure called singular value decomposition (Hardeberg, 2001). Imagine an $n \times w$ matrix $\mathbf{P}$ that contains n spectra each sampled at w wavelengths. Singular value decomposition decomposes the matrix $\mathbf{P}$ thus,

$$\mathbf{P} = \mathbf{UWV}^{\mathrm{T}}, \tag{10.8}$$

where $\mathbf{U}$ and $\mathbf{V}$ are $n \times n$ and $w \times w$ matrices, respectively. The matrix $\mathbf{W}$ is an $n \times w$ matrix where diagonal entries denote singular values of $\mathbf{P}$ (Pratt, 1978). The columns of $\mathbf{U}$ are the eigenvectors of the matrix $\mathbf{PP}^{\mathrm{T}}$ and these may be used as the basis functions. Similarly, the columns of $\mathbf{V}$ are the eigenvectors of $\mathbf{P}^{\mathrm{T}}\mathbf{P}$. Computer code (in the C programming language) to perform a singular value decomposition of a matrix is readily available (e.g. Press *et al.*, 1993). MATLAB provides the commands *svd* and *svds* which can be readily used to generate the eigenvectors for a set of data.

Strictly speaking, for Principal Components Analysis (PCA), the mean of $\mathbf{P}$ should subtracted from $\mathbf{P}$ to yield a new matrix and it is the eigenvectors from the singular value decomposition of this matrix that yield the principal components. In practice, however, it is not always necessary to subtract the mean to use a linear model of reflectance. If we use a linear model to represent a set of reflectance spectra, then a given sample in the set is given by the linear sum of the basis functions weighted by coefficients so that

$$P(\lambda) = a_1 B_1(\lambda) + a_2 B_2(\lambda) + a_3 B_3(\lambda) \ldots a_n B_n(\lambda), \tag{10.9}$$

and if all n basis functions are used all the spectra in the set can be reconstructed perfectly using appropriate values of the weights $a_1 \ldots a_n$. However, the benefit of techniques such as PCA is that it is possible to represent data efficiently by only using a small number of basis functions. The first basis function maximally represents the variance in the data, and subsequent basis functions maximally represent the remaining variance. It has been shown that more than 95% of the variance in a set of reflectance spectra can be represented by using just the first three basis functions (Maloney, 1986; Owens, 2002b). Figure 10.2 shows the first

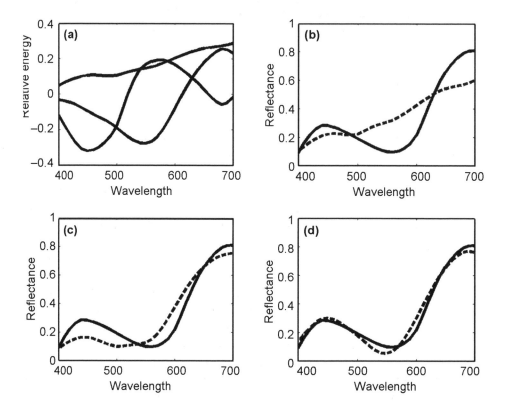

Figure 10.2 (a) First three basis functions in a linear model of reflectances; a typical reflectance spectrum (solid line) approximated by one (b), two (c) and three (d) basis functions from (a)

three basis functions (also known as eigenvectors) computed for a set of reflectance spectra measured from a CMYK printing process. The figure shows the approximation of one of the spectra by one, two and three basis functions. When three basis functions are used the approximation is quite a good fit to the measured values.

It is not trivial, however, to ascertain how many basis functions are required for an accurate representation without reference to the purpose of the representation. Owens (2002b) measured the reflectance spectra of a set of natural surfaces collected from the grounds of Keele University and compared these with a set of Munsell reflectance spectra. Figure 10.3(a) shows how the mean-squared error for the two sets monotonically decreases with the number of basis functions if a set of basis functions derived from the Keele data is used to represent the Keele data and a set of basis functions from the Munsell data is used to represent the Munsell data. In Figure 10.3(b), however, it can be seen that at least six basis functions are required if average CIELAB ΔE values of about 1 are required.

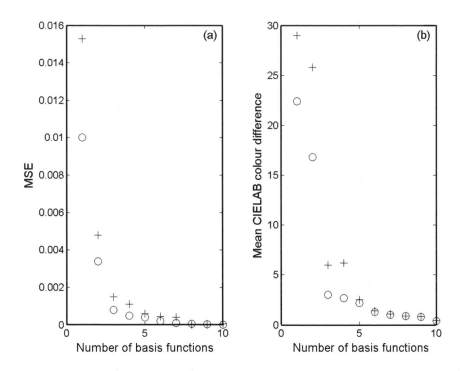

Figure 10.3   (a) Plot of the mean-square error (MSE) as a function of number of basis functions for Keele ($\circ$) and Munsell ($+$) data; (b) plot of mean CIELAB ΔE as a function of number of basis functions for Keele ($\circ$) and Munsell ($+$) data

10.3   Surface and illuminant estimation algorithms

Most algorithms for reflectance estimation rely on two essential components (Brill, 1979). First, we need a method of representing our knowledge about the likely surface and illuminant functions (for example, linear models). Secondly, most modern estimation methods assume that the illumination varies either slowly or not at all across the image. This assumption is important because it means that the illumination adds very few extra parameters that need to be estimated.

Consider an image with p distinct spatial positions. We expect to obtain three cone excitations at each position in the image so the number of measurements is $3p$ in total. If we can use a three-dimensional model for the surfaces, then there are a total of $3p$ unknown surface coefficients. If the illuminant is known, then the problem is easy to solve since we have as many measurements ($3p$) as unknowns ($3p$). If the illuminant is not known and can vary from point to point, then there will be $6p$ unknown parameters (at each point, three parameters for the surface and three parameters for the illuminant) and the problem cannot be solved. If the illuminant is constant across the image we have only three additional parameters (thus $3p + 3$ unknowns and $3p$ measurements) and by making some modest assumptions we can proceed with the estimation algorithm.

Modern estimation algorithms work by finding a method to overcome the mismatch between the measurements and the unknowns. The majority of algorithms infer the illumination parameters by making one additional assumption about the image contents. For example, if we know the reflectance function of just one object in the scene, then we can use the sensor responses from that object to estimate the illuminant. This is often implemented in terms of the assumption that the average of all the surfaces in the image is grey – the so-called grey-world assumption (Land, 1986; Wandell, 1995). Other algorithms are based on the assumption that the brightest surface in the image is a uniform perfect reflector (Wandell, 1995). Another interesting idea is that we can identify specularities in the image from glossy surfaces (D'Zmura and Lennie, 1986; Tominaga and Wandell, 1990).

A second group of estimation algorithms compensates for the mismatch in measurements and parameters by suggesting ways to acquire more data. Maloney and Wandell (1986) showed that by adding a fourth sensor one can estimate the surface and illuminant. Similarly, D'Zmura and Iverson (1993a, 1993b) explored the possibility of observing the same surface under different illuminants. However, even if the illuminant is known the number of unknowns may be greater than $3p$ if it is assumed that a linear model with greater than three dimensions is required to represent the reflectance spectra. Multispectral imaging is a technique that uses more than three channels so that sufficient information about the scene is captured to allow spectral recovery to an accuracy greater than that which would be possible using a three-dimensional linear model.

10.4   Techniques for multispectral imaging

In this section we consider some typical techniques to allow reflectance recovery using multispectral imaging.

10.4.1   The Hardeberg method

The method proposed by Hardeberg (1999) assumes a linear camera model [Equation (10.1)] and has similarities to the method of reflectance recovery proposed by Maloney and Wandell (1986). For a single surface, the Hardeberg method is based upon Equation (10.10), so that

$$\mathbf{r} = \Lambda \mathbf{a}, \tag{10.10}$$

where $\mathbf{r}$ is an $r \times 1$ matrix of sensor responses, Λ is an $r \times n$ system matrix and $\mathbf{a}$ is an $n \times 1$ column matrix of weights that defines the surface in the space of basis functions. Since it is known (Maloney, 1986; Owens, 2002b) that a linear model of at least six basis functions is required for the accurate representation of reflectance spectra, $\mathbf{r}$ must be at least size 6×1. Most practical multispectral imaging systems therefore consist of at least six separate channels and this may be achieved by a filter wheel containing a number of different filters and a monochrome camera system.

If we consider the case where $r = n = 3$, then the 3×3 matrix Λ would be obtained from the product of the 3×31 matrix of the sensor spectral sensitivities (weighted by the illuminant power distribution) and the 31×3 matrix of the basis functions of the linear model. The entries of Λ are thus known.

The reflectance of the surface may be recovered by manipulating Equation (10.10) to yield

$$\mathbf{a} = \Lambda^{-1} \mathbf{r}, \tag{10.11}$$

where the inverse Λ^{-1} may easily be computed if $r = n$ (when Λ is a square matrix). An alternative procedure to using more sensor classes is to use more than one light source. For example, if an image is taken using a trichromatic camera using one light source and then the same image is taken using a second light source, then it allows the construction of a matrix Λ with six rows and consequently allows a linear model of reflectance spectra with six basis functions. It is important, however, that the spectral power distributions of the two or more light sources are as unrelated as possible and that they contain power throughout the whole visible spectrum (Connah et al., 2001).

Hardeberg (1999) has also considered the situation where $r > n$ so that the number of sensor classes exceeds the number of basis functions in the linear model of reflectance. This leads to Equation (10.12),

$$\mathbf{a} = \Lambda^{+} \mathbf{r}, \tag{10.12}$$

where the pseudoinverse of the non-square matrix $\mathbf{\Lambda}$ is computed. Equation (10.12) refers to an over-determined system, but this technique can actually lead to improved estimates when compared with the case $r = n$. There are two reasons why an over-determined system may be useful. First, for a system based upon r sensors the r rows of $\mathbf{\Lambda}$ may not be independent. This can happen if the spectral sensitivities of the channels are correlated with each other (or, for a system using two light sources, if the spectral power distributions of the light sources are correlated). Secondly, estimates of $\mathbf{a}$ when $r = n$ may suffer if the system is noisy so that the matrix $\mathbf{r}$ is known with low precision.

10.4.2 The Imai and Berns method

Imai and Berns (1999) have developed a method for reflectance recovery based directly upon Equations (10.10) and (10.11). They assume a linear relationship between the sensor outputs $\mathbf{r}$ of the imaging system and the representation of the surface in an r-dimensional basis space by the weights $\mathbf{a}$. However, unlike the Hardeberg method, Imai and Berns find the entries of $\mathbf{\Lambda}$ using an empirical least-squares analysis. The method is simple and effective because for the Hardeberg method it is necessary to determine the space of basis functions in which the reflectance spectra will be represented, to measure the spectral power distribution of the light source and to determine the spectral sensitivities of the imaging system. The method proposed by Imai and Berns, however, requires only the first of these steps, namely the determination of the basis functions, and the entries of $\mathbf{\Lambda}$ are then found by optimization.

10.4.3 Methods based on maximum smoothness

One problem with methods for reflectance recovery that use basis functions is that the recovered reflectance cannot be guaranteed to be within the range [0, 1]. The methods described in Sections 10.4.1 and 10.4.2 do not always yield physically reasonable solutions. An alternative approach to reflectance recovery is to replace the constraint imposed by the linear model of basis functions with some other constraint. One possibility is to employ a constraint of maximum smoothness (Li and Luo, 2001).

10.5 Implementations and examples

10.5.1 Deriving a set of basis functions

Principal Component Analysis (PCA) may be performed using MATLAB's singular value decomposition function *svds*. Consider the 100 observations of two variables x and y as illustrated in Figure 10.4. The principal components of these data are obtained by creating what is called a centred matrix (by subtracting from each observation the mean x and y values) and then using the *svds* command. The *svds* function can be called with two arguments where the first argument is the matrix of data (with the number of samples along the rows and the dimensions along the column) and the second argument is the number of basis functions or eigenvectors that are computed. So, for example, the code

```
load xydata.mat
% x is a 100 × 1 matrix
% y is a 100 × 1 matrix
cx = x - mean(x);
cy = y - mean(y);
data = [cx cy];
[u,s,v] = svds(data,2);
```

results in the 2×2 matrix **V** which, for the data in Figure 10.4, has the values shown in Equation (10.13):

$$\mathbf{V} = \begin{bmatrix} 0.3655 & -0.9308 \\ 0.9308 & 0.3655 \end{bmatrix},$$
(10.13)

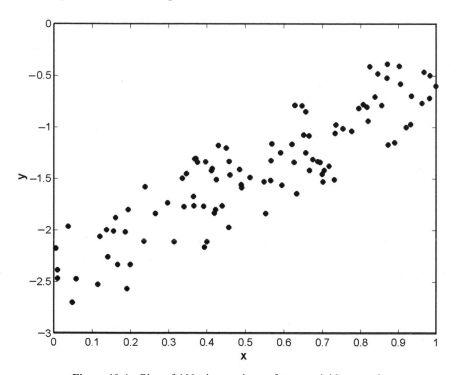

Figure 10.4 Plot of 100 observations of two variables x and y

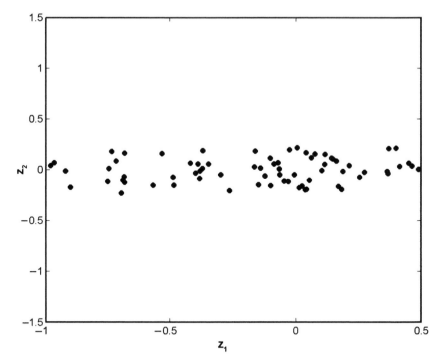

Figure 10.5 The centred data from Figure 10.4 are redrawn along the axes z_1 and z_2

where the first column of **V** represents the first component and the second column represents the second component. We can use these components to create two new axes, z_1 and z_2, where

$$z_1 = 0.3655x + 0.9308y,$$
$$z_2 = -0.9308x + 0.3655y.$$
(10.14)

The MATLAB code

```
tdata = v'*data';
```

transforms the xy data in the data matrix into the dimensions of z_1 and z_2. The data in Figure 10.4 are redrawn in Figure 10.5 using the new orthogonal axes z_1 and z_2 from which it is clear that the new axes more appropriately describe the variation in the data set.

The basis functions that describe a particular set of reflectance spectra can similarly be obtained using MATLAB's singular value decomposition function *svds*. The following MATLAB code has been used to generate the basis functions for a set of 404 reflectance spectra using two methods and to generate Figure 10.6.

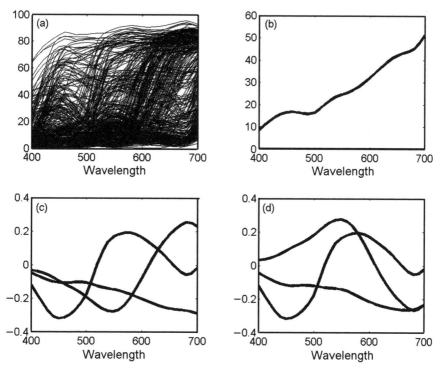

Figure 10.6 Computation of basis functions: (a) the set of 404 reflectance spectra; (b) average reflectance spectrum of the set; (c) basis functions derived without subtracting the mean from the set; (d) basis functions derived after subtracting the mean from the set

```
clear
load spectra.txt
% spectra is a 404 by 31 matrix
w=linspace(400,700,31);
% compute the mean spectrum
mspec = mean(spectra);
% create a new centred matrix
dspec = spectra;
for i=1:404
   dspec(i,:) = spectra(i,:)-mspec;
end
% compute the first three basis functions
[u,s,v]=svds(spectra,3);
[u,s,v1]=svds(dspec,3);
% generate the plots
subplot(2,2,1)
plot(w,spectra,'k-')
```

```
xlabel('wavelength')
title('(a)')
subplot(2,2,2)
plot(w,mspec,'k-')
xlabel('wavelength')
title('(b)')
subplot(2,2,3)
plot(w,v,'k-')
xlabel('wavelength')
title('(c)')
subplot(2,2,4)
plot(w,v1,'k-')
xlabel('wavelength')
title('(d)')
```

The lower left (c) and lower right (d) panes of Figure 10.6 show the first three basis functions computed from the raw set of reflectance spectra and from a centred set of spectra (where the mean is first subtracted), respectively. Although the two sets of basis functions look quite different if we correct for the arbitrary sign of the functions it can be seen that there are only small differences between the two sets of basis functions (Figure 10.7).

10.5.2 Representation of reflectance spectra in a linear model

The computation of the basis functions using *svds* allows us to write

$$\mathbf{P} = \mathbf{Ba}, \tag{10.15}$$

where $\mathbf{P}$ is the $w \times n$ matrix of reflectance spectra, $\mathbf{B}$ is the $w \times m$ matrix of basis functions, and $\mathbf{a}$ is the $m \times n$ matrix of coefficients, where n is the number of samples, w is the number of wavelength intervals at which the samples are represented and m is the number of basis functions in the linear model. The coefficient matrix $\mathbf{a}$ thus allows each reflectance spectrum to be represented by just m coefficients. The central goal of PCA is to reduce the dimensionality of a data set whilst retaining as much as possible of the variation present in the data set (Jolliffe, 1986). It is relatively straightforward to compute the coefficient matrix $\mathbf{a}$ by rearranging Equation (10.16),

$$\mathbf{a} = \mathbf{B}^+ \mathbf{P}, \tag{10.16}$$

where $\mathbf{B}^+$ denotes the pseudoinverse of the matrix of basis functions $\mathbf{B}$. We note, however, that if the basis functions are orthonormal, then Equation (10.16) is equivalent to Equation (10.17),

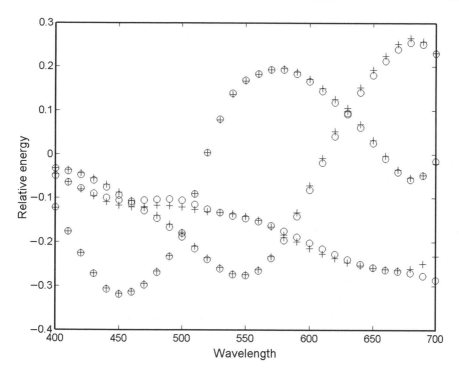

Figure 10.7 The basis functions computed without subtracting the mean reflectance ($\circ$) are compared with the basis functions computed after subtracting the mean reflectance ($+$)

$$\mathbf{a} = \mathbf{B}^{\mathrm{T}}\mathbf{P}, \qquad (10.17)$$

where $\mathbf{B}^{\mathrm{T}}$ denotes the transpose of the matrix $\mathbf{B}$. A set of vectors is called an orthogonal set if all pairs of distinct vectors in the set are orthogonal. An orthogonal set in which each vector has norm 1 is called orthonormal (Anton, 1994). Two non-zero vectors are orthogonal if and only if their dot product is zero. If $\mathbf{b}_1$ is a $1 \times w$ row matrix representing the first basis function and $\mathbf{b}_2$ is a $w \times 1$ column matrix representing the second basis function, then we can say that $\mathbf{b}_1$ and $\mathbf{b}_2$ are orthogonal if Equation (10.18) is satisfied,

$$\mathbf{b}_1\mathbf{b}_2 = 0. \qquad (10.18)$$

The norm of a matrix can be computed by the MATLAB function *norm*. The norm of a matrix is also called the length of the matrix. Matrix $\mathbf{b}_1$ will be of length 1 if Equation (10.19) is satisfied,

$$\mathbf{b}_1^{\mathrm{T}}\mathbf{b}_1 = 1. \qquad (10.19)$$

The special property of orthonormality allows Equation (10.17) to be used instead of Equation (10.16) because the transpose of a matrix of length 1 is equal to its inverse. Equation (10.17) is clearly easier to implement in a programming

Table 10.1 Reconstruction errors for linear models using six basis functions

Method	Median ΔE	Maximum ΔE
Without subtracting the mean	0.89	13.24
After subtracting the mean	0.81	12.58

language such as C than Equation (10.16) which requires the computation of a pseudoinverse.

The following code has been used to implement linear models using six basis functions for the case where the mean is subtracted from a set of 404 reflectance spectra and for the case where the mean is not subtracted. The reconstructed spectra have been computed and the CIELAB ΔE values have been calculated between the target and reconstructed spectra (Table 10.1). It is evident from Table 10.1 that lower reconstruction errors result when the mean is subtracted from the spectra before using the *svds* command but that the difference between the two methods is quite small. Note, however, that for other data sets it may be possible to find the opposite effect, namely that the basis functions computed without subtracting the mean from the data generate the lower ΔE values:

```
clear
load spectra.txt
% spectra is a 404 by 31 matrix of reflectance values

mspec = mean(spectra);
dspec = spectra;
for i=1:404
  dspec(i,:) = spectra(i,:)-mspec;
end
[u,s,v]=svds(spectra,6);
[u,s,v1]=svds(dspec,6);

spectra=spectra';
dspec = dspec';

% compute the coefficients in basis space
a = pinv(v)*spectra;
a1 = pinv(v1)*dspec;

% reconstruct the spectra from the basis functions
pspectra = v*a;
pdspec = v1*a1;
```

```
for i=1:404
  pdspec(:,i) = pdspec(:,i)+mspec';
end

% compute the reconstruction errors
de1 = zeros(404,1); de2 = zeros(404,1);
for i=1:404
  xyzt = r2xyz(spectra(:,i),400,700,'d65_64');
  xyz1 = r2xyz(pspectra(:,i),400,700,'d65_64');
  xyz2 = r2xyz(pdspec(:,i),400,700,'d65_64');
  labt = xyz2lab(xyzt,'d65_64');
  lab1 = xyz2lab(xyz1,'d65_64');
  lab2 = xyz2lab(xyz2,'d65_64');
  thisde1 = cielabde(labt,lab1);
  thisde2 = cielabde(labt,lab2);
  de1(i) = thisde1;
  de2(i) = thisde2;
end

result = [median(de1) max(de1) median(de2) max(de2)]
```

10.5.3 Estimation of reflectance spectra from tristimulus values

Despite the fact that spectral reflectance factors are almost always smooth functions of wavelength and are highly constrained it is not possible to accurately compute reflectance spectra from tristimulus values. Clearly, since metamerism exists, the mapping from $\mathbf{T} \rightarrow \mathbf{P}$, where $\mathbf{T}$ is a $3 \times n$ matrix of tristimulus values and $\mathbf{P}$ is a $31 \times n$ matrix of reflectance values, is a one-to-many mapping. However, the use of linear models and basis functions allows the estimation of a possible reflectance spectrum corresponding to a target triplet of tristimulus values.

We can represent the computation of the tristimulus values $\mathbf{t}$ for a given reflectance spectrum $\mathbf{p}$ by the linear system

$$\mathbf{t} = \mathbf{Mp}, \tag{10.20}$$

where $\mathbf{M}$ is a 3×31 matrix whose rows contain the wavelength-by-wavelength product of the illuminant with one of the three colour-matching functions. We can try to solve Equation (10.20) directly by rearranging to give

$$\mathbf{p} = \mathbf{M^+t}, \tag{10.21}$$

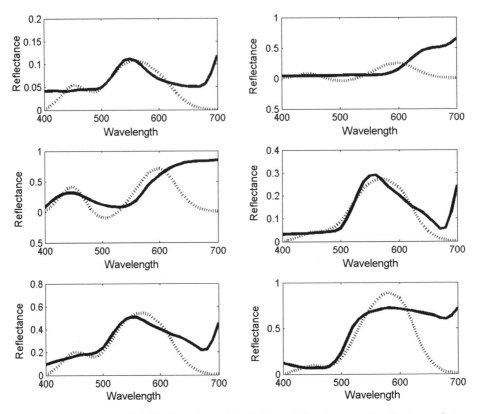

Figure 10.8 Target (solid lines) and predicted (dotted lines) spectral reflectance factors computed using Equation (10.21) for six samples

but estimates of **p** are likely to be wildly inaccurate. Figure 10.8 shows the predicted reflectance spectra for six samples using this method.

If we use the a priori knowledge of the smoothness of reflectance spectra, then the problem may be better constrained and more accurate predictions may be possible. The a priori knowledge is represented by the basis functions. So, for example, if we use a linear model with three basis functions, then the $31 \times n$ matrix of reflectance spectra **P** in Equation (10.20) can be replaced by **Ba**, where **B** is a 31×3 matrix of basis functions and **A** is a $3 \times n$ matrix of coefficients to produce Equation (10.22),

$$\mathbf{T} = \mathbf{MBA}. \tag{10.22}$$

MB is a 3×3 matrix and therefore Equation (10.22) now represents a linear system with three constraints and three unknowns and can be easily solved using Equation (10.23),

$$\mathbf{A} = (\mathbf{MB})^{-1}\mathbf{T}. \tag{10.23}$$

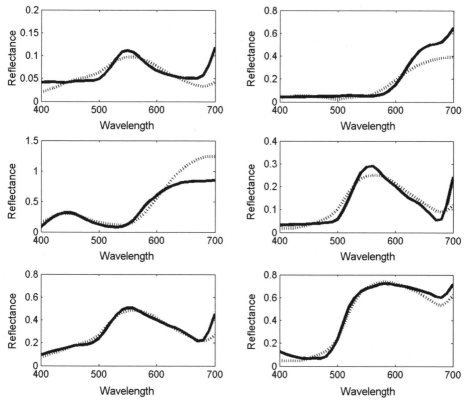

Figure 10.9 Target (solid lines) and predicted (dotted lines) spectral reflectance factors computed using Equation (10.24) for six samples

Now, since $\mathbf{P} = \mathbf{BA}$, we can write an equation to recover the reflectance spectra thus,

$$\mathbf{P} = \mathbf{B}(\mathbf{MB})^{-1}\mathbf{T}. \tag{10.24}$$

Figure 10.9 shows the predicted reflectance spectra using this method for the same six samples as are shown in Figure 10.8. The basis functions were computed from a set of 404 samples that contained these six samples. The accuracy of the reconstructed spectra in Figure 10.9 is much improved compared with those in Figure 10.8 and all of the predicted reflectance spectra are within the range [0, 1]. However, if we consider all 404 samples, 82 still contain predicted reflectance factors outside of the range [compared with 129 samples for the method based upon Equation (10.21)].

Clearly, additional constraints are necessary if this method is to generate physically reasonable reflectance factors in all cases. However, a function called *xyz2r* has been provided based upon this method. A typical call would be

```
[P] = xyz2r(XYZ, obs);
```

where, the matrix **XYZ** is an $n \times 3$ matrix of tristimulus values and **obs** is a string that defines the illuminant/observer combination (e.g. 'd65_64'). A matrix **P** is returned with the reflectance values with size $31 \times n$.

Box 26: *xyz2r.m*

```
function [P] = xyz2r(XYZ, obs)

% function [P] = xyz2r(XYZ, obs)
% estimates reflectance P from XYZ trimulus values
% matrix XYZ must be n by 3
% illuminants A, C, D50, D55, D65, D75, F2, F7, F9
% set obs to 'd65_64 for D65 and 1964, for example
% matrix P is returned as an n by 31 matrix

% check dimensions of XYZ
dim = size(XYZ);
if (dim(2) ~= 3)
  disp('XYZ must be n by 3');
  return;
end

load evectors.mat
% loads the 31 by 3 matrix v containing three basis
functions

load weights % contains the tables of weights
if strcmp('a_64', obs)
  cie = a_64;
elseif strcmp('a_31', obs)
  cie = a_31;
elseif strcmp('c_64', obs)
  cie = a_64;
elseif strcmp('c_31', obs)
  cie = c_31;
elseif strcmp('d50_64', obs)
  cie = d50_64;
elseif strcmp('d_50', obs)
  cie = d_50;
elseif strcmp('d55_64', obs)
  cie = d55_64;
```

```
elseif strcmp('d55_31', obs)
  cie = d55_31;
elseif strcmp('d65_64', obs)
  cie = d65_64;
elseif strcmp('d65_31', obs)
  cie = d65_31;
elseif strcmp('d75_64', obs)
  cie = d75_64;
elseif strcmp('d75_31', obs)
  cie = d75_31;
elseif strcmp('f2_64', obs)
  cie = f2_64;
elseif strcmp('f2_31', obs)
  cie = f2_31;
elseif strcmp('f7_64', obs)
  cie = f7_64;
elseif strcmp('f7_31', obs)
  cie = f7_31;
elseif strcmp('f9_64', obs)
  cie = f9_64;
elseif strcmp('f9_31', obs)
  cie = f9_31;
else
  disp('unknown option obs');
  disp('use d65_64 for D65 and 1964 observer'); return;
end

% the basis functions are only available in the range
400-700 nm
M = cie(5:35,:);
M(1,:) = M(1,:) + sum(cie(1:4,:));
M(31,:) = M(31,:) + sum(cie(36:43,:));

Q = v'*M
a = XYZ*inv(v'*M)
P = a*v';
```

10.5.4 Estimation of reflectance spectra from camera responses

The following MATLAB code implements the method of Imai and Berns (1999) to predict reflectance spectra from camera *RGB* responses. The data that were introduced in Chapter 8 (Cheung and Westland, 2004) were used for this analysis. Thus *RGB* values were available from the Agfa StudioCam camera for 192 samples from the Macbeth DC Colorchecker (the training samples) and the 24 samples of the Macbeth Colorchecker (the test samples). The *RGB* values were linearized and corrected for spatial non-uniformity of the device and illumination (see Chapter 8) and were saved along with the reflectance spectra for each of the samples in a file called *agfa.mat*.

```
clear
load agfa.mat
% refldc is a 192 by 31 matrix of reflectance
% reflck is a 24 by 31 matrix of reflectance
% rgbdc is a 192 by 3 matrix of RGB values
% rgbck is a 24 by 3 matrix of RGB values

% compute the basis functions from the Macbeth DC samples
[u,s,v] = svds(refldc,3);

% compute the matrix of weights for the Macbeth DC samples
adc = pinv(v)*refldc';

% assume that T=Ma and solve for M
M = rgbdc'*pinv(adc);

% use the linear transform M to predict the weights
% from the camera values
padc = inv(M)*rgbdc';
pack = inv(M)*rgbck';

% now reconstruct the spectra
prefldc = v*padc;
prefldc = prefldc';
preflck = v*pack;
preflck = preflck';

% compute the colour differences for the reconstructions
deck = zeros(24,1);
dedc = zeros(192,1);
for i=1:192
  xyzt = r2xyz(refldc(i,:),400,700,'d65_64');
  xyzp = r2xyz(prefldc(i,:),400,700,'d65_64');
  labt = xyz2lab(xyzt,'d65_64');
  labp = xyz2lab(xyzp,'d65_64');
```

```
de = cielabde(labt,labp);
  dedc(i) = de;
end

for i=1:24
  xyzt = r2xyz(reflck(i,:),400,700,'d65_64');
  xyzp = r2xyz(preflck(i,:),400,700,'d65_64');
  labt = xyz2lab(xyzt,'d65_64');
  labp = xyz2lab(xyzp,'d65_64');
  de = cielabde(labt,labp);
  deck(i) = de;
end
```

10.5.5 Fourier operations on reflectance spectra

The Fourier properties of reflectance spectra may be computed using the MATLAB function *fft*. For a more complete description of Fourier analysis of discrete signals using MATLAB, the reader is directed towards the text by Carlson (1998). The *fft* command decomposes a signal into its frequency and phase components. Equation 10.25 shows a hypothetical reflectance function P that is a function of wavelength λ,

$$P(\lambda) = b + A\cos(2\pi f\lambda + \phi). \tag{10.25}$$

The signal P may be represented entirely by the value of its offset b, its amplitude A, its frequency f and its phase ϕ. The frequency may be represented in terms of cycles per nanometer. So, for example, if $b = 0.5$, $A = 0.5$, $f = 0.005$ cyc/nm and $\phi = 0$, then we would obtained the signal shown (between the wavelengths 360 and 780 nm) by Figure 10.10.

The spectrum shown in Figure 10.10 consists of a single spectral frequency of 0.005 cyc/nm or (0.005)*300 = 1.5 cycles in the visible spectrum (400–700 nm). Fourier analysis involves taking a signal (such as a reflectance curve) and decomposing it into an amplitude spectrum and a phase spectrum. The amplitude spectrum provides information about the spectral frequencies that are present in the reflectance data and the phase spectrum provides information about the phases of these components. If the data are band limited, then there will be a limiting spectral frequency (known as the band limit) above which there is no further energy. The amplitude and frequency information can be obtained using the following two MATLAB commands,

```
four_amp = abs(fft(p));
four_phase = angle(fft(p));
```

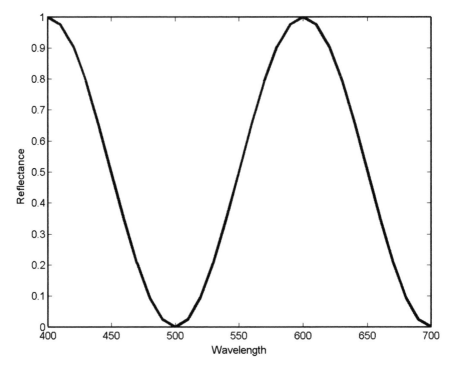

Figure 10.10   Hypothetical reflectance curve with a single spectral frequency

where **p** is a $1 \times w$ row matrix and w is the number of wavelength intervals at which the signal is represented. If **p** is a 1×31 row matrix (representing reflectance data at 10-nm intervals in the range 400–700 nm), then **four_amp** and **four_phase** will also be 1×31 row matrices. Figure 10.11 shows a typical reflectance specrum [Figure 10.11(a)] and the 31-dimensional vector four_amp that results. The first 16 values are the amplitudes at evenly spaced spectral frequencies between zero (also known as the DC) and the Nyquist limit (see Carlson, 1998). The Nyquist limit is half the sampling rate and since the sampling rate is 31, the Nyquist limit is 31/2 cycles per 300 nm or 0.0517 cyc/nm. The subsequent 15 values are the amplitudes for the negative frequencies; these are a mirror-image of those for the positive frequencies and are thus often discarded to create the plots shown in Figure 10.11(c) and also Figure 10.11(d).

Figure 10.11(c) shows that the amplitude generally falls off as the spectral frequency increases, indicating that the reflectance data are band limited. Apart from the DC component, Figure 10.11(c) also shows that the frequency with the greatest amplitude is 0.0034 cyc/nm, which corresponds to about a single complete cycle in the range 400–700 nm.

The Fourier representation is useful for analysing properties of reflectance spectra and also for performing various operations such as smoothing (convolution in the wavelength domain may be achieved by a multiplication in

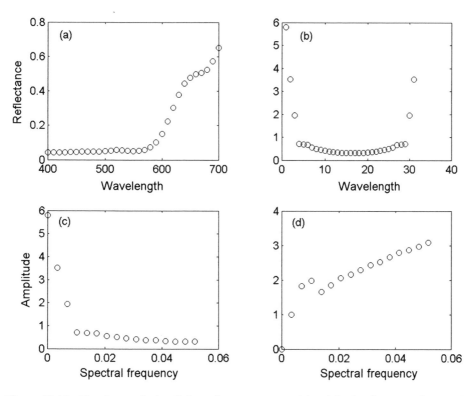

Figure 10.11   Fourier analysis of the reflectance curve: (a) original reflectance data at 31 wavelength intervals; (b) 31×1 vector of amplitude information; (c) plot of amplitude vs. positive frequencies; (d) plot of phase vs. positive frequencies

the Fourier domain). Following operations upon the matrices **four_amp** and **four_phase**, the corresponding data in the reflectance domain may be obtained using the commands

```
four_x = four_amp.*cos(four_phase);
four_y = four_amp.*sin(four_phase);
rec_p = real(ifft(complex(four_x,four_y)));
```

11

Colour Toolbox

The following list details the functions that have been written as part of this book. These programs form a computational toolbox. All these functions can be downloaded from http://www.colourware.co.uk/compute/ and from http://www.mathworks.com/matlabcentral/. To install the toolbox we recommend that these files be placed in a directory called colour in MATLAB's toolbox directory. For example, this may be c:\Program Files\MATLABp65\ toolbox\colour\ but the exact pathname will depend upon how MATLAB was installed. If the colour directory is then added to MATLAB's environment path, then the colour toolbox will be available as any other toolbox.

11.1 *cband.m* (Box 1)

```
% function [cP] = cband(P)
% applies Stearns Stearns spectral bandpass correction
% operates on matrix P of dimensions 1 by n
% where n is the number of wavelengths
% returns corrected matrix cP
```

11.2 *pinterp.m* (Box 2)

```
% function [s] = pinterp(p)
% applies interpolation to double the sampling
% rate of the n by 1 matrix p
% returns interpolated matrix s
```

Computational Colour Science Using MATLAB. By Stephen Westland and Caterina Ripamonti.
© 2004 John Wiley & Sons, Ltd: ISBN 0 470 84562 7

11.3 *r2xyz.m* (Box 3)

```
% function [xyz] = r2xyz(p, startlam, endlam, obs)
% computes XYZ from reflectance p using a table of weights
% operates on matrix p of dimensions 1 by n for
% illuminants A, C, D50, D55, D65, D75, F2, F7, F9
% and for the 1931 and 1964 observers
% set obs to 'd65_64 for D65 and 1964, for example
% the startlam and endlam variables denote the first and
% last wavelengths (eg. 400 and 700) for your reflectance
% which must be integers of 10 in the range 360-780
```

11.4 *plocus.m* (Box 4)

```
% function [xy] = plocus(obs)
% computes spectral locus xy using interpolated ASTM
% weights
% see function r2xyz for valid values for obs
```

11.5 *xyz2lab.m* (Box 5)

```
% function [lab] = xyz2lab(xyz,obs)
% computes CIELAB LAB values from XYZ tristimulus values
% requires the illuminant/observer obs to define white
% point
% see function r2xyz for valid values for obs
```

11.6 *lab2xyz.m* (Box 6)

```
% function [xyz] = lab2xyz(lab,obs)
% computes XYZ tristimulus values from CIELAB LAB values
% requires the illuminant/observer obs to define white
% point
% see function r2xyz for valid values for obs
```

11.7 *xyz2luv.m* (Box 7)

```
% function [luv,uprime,vprime] = xyz2luv(xyz,obs)
% computes CIELUV Luv values from XYZ tristimulus values
% uprime and vprime are the CIE 1976 UCS coordinates
% requires the illuminant/observer obs to define white
% point
% see function r2xyz for valid values for obs
```

11.8 *car2pol.m* (Box 8)

```
% function [c,h] = cartopol(ab)
% converts a*b* or u*v* into the polar coordinates
% of Chroma C and Hue H
% ab must be a row or column matrix 2 by 1 or 1 by 2
% see also pol2car
```

11.9 *pol2car* (Box 9)

```
% function [a,b] = pol2car(ch)
% converts the polar coordinates
% of Chroma C and Hue H
% ch must be a row or column matrix 2 by 1 or 1 by 2
% see also car2pol
```

11.10 *cielabde.m* (Box 10)

```
% function [de,dl,dc,dh] = cielabde(lab1,lab2)
% computes colour difference from CIELAB values
% using CIELAB formula
% lab1 and lab2 must be 3 by 1 or 1 by 3 matrices
% and contain L*, a* and b* values
% see also cmcde, cie94de, and cie00de
```

11.11   *dhpolarity* (Box 11)

```
% function [p] = dhpolarity(lab1,lab2)
% computes polarity of hue difference
% lab1 and lab2 must be 3 by 1 or 1 by 3 matrices
% and contain L*, a* and b* values
% p is +1 if the hue of the trial (lab2) is anticlockwise
% from the standard (lab1) and -1 otherwise
```

11.12   *cmcde.m* (Box 12)

```
% function [de,dl,dc,dh] = cmcde(lab1,lab2,paral,parac)
% computes colour difference from CIELAB values
% using CMC(l:c) formula
% lab1 and lab2 must be 3 by 1 or 1 by 3 matrices
% and contain L*, a* and b* values
% The dl, dc and dh components are CMC deltas
% The defaults for paral and parac are 1
% see also cielabde, cie94de, and cie00de
```

11.13   *cie94de.m* (Box 13)

```
% function [de,dl,dc,dh] = cie94de(lab1,lab2)
% computes colour difference from CIELAB values
% using the CIE94 formula
% lab1 and lab2 must be 3 by 1 or 1 by 3 matrices
% and contain L*, a* and b* values
% The dl, dc and dh components are CIE94 deltas
% see also cielabde, cmcde, and cie00de
```

11.14 *cie00de.m* (Box 14)

```
% function [de,dl,dc,dh] = cie00de(lab1,lab2,paral,
parac,parah)
% computes colour difference from CIELAB values
% using the CIEDE2000 formula
% lab1 and lab2 must be 3 by 1 or 1 by 3 matrices
% and contain L*, a* and b* values
% The dl, dc and dh components are CIEDE2000 deltas
% The defaults for paral, parac and parah are 1
% see also cielabde, cmcde, and cie94de
```

11.15 *cmccat97.m* (Box 15)

```
% function [xyzc] = cmccat97(xyz,xyzt,xyzr,la,f)
% implements the CMCCAT97 chromatic adaptation transform
% operates on 1 by 3 matrix xyz containing tristimulus
% values of the stimulus under the test illuminant
% xyzt and xyzr are 1 by 3 matrices containing the
% white points for the test and reference conditions
% f has default value 1
% la is the luminance of the adapting test field
% and has default value of 100
% xyzc contains the tristimulus values of the
% stimulus under the reference illuminant
% see also cmccat00
```

11.16 *cmccat00.m* (Box 16)

```
% function [xyzc] = cmccat00(xyz,xyzt,xyzr,lt,lw,f)
% implements CMCCAT2000 chromatic adaptation transform
% operates on 1 by 3 matrix xyz containing tristimulus
% values of the stimulus under the test illuminant
% xyzt and xyzr are 1 by 3 matrices containing the
% white points for the test and reference conditions
% f has default value 1
```

```
% lt is the luminance of the adapting test field
% and has default value of 100
% lw is the luminance of the adapting reference field
% and has default value of 100
% xyzc contains the tristimulus values of the
% stimulus under the reference illuminant
```

11.17   *ciecam97s.m* (Box 17)

```
% function [j,c,hq,m,h,s,q,cd]=ciecam97s(xyz,xyzw,
la,yb,para)
% implements the CIECAM97s colour appearance model
% operates on 1 by 3 matrix xyz containing tristimulus
% values of the stimulus under the test illuminant
% xyzt and xyzr are 1 by 3 matrices containing the
% white points for the test and reference conditions
% la and yb are the luminance and Y tristimulus values of
% the achromatic background against which the sample is
% viewed
% para is a 1 by 4 matrix containing c, Nc, Fll and F
```

11.18   *gogtest.m* (Box 18)

```
% function [err] = gogtest(gogs,dacs,rgbs)
% computes the error between measured and predicted
% linearized dac values for a given set of GOG values
% gogs is a 2 by 1 matrix that contains the gamma and gain
% dacs is an n by 1 matrix that contains the actual RGB values
% obtained by dividing the RGB values by 255
% rgbs is an n by 1 matrix that is obtained from a linear
% transform of measured XYZ values
```

11.19   *compgog.m* (Box 19)

```
% function [rgb] = compgog(gogs,dacs)
% computes the linearized RGB values
% from the normalized RGB values
% for a given set of gog values
% gog is a 2 by 1 matrix that contains the gamma and gain
% dacs is an n by 1 matrix that contains the RGB values
% rgb is an n by 1 matrix of linearized RGB values
```

11.20   *rgb2xyz.m* (Box 20)

```
% function [XYZ] = rgb2xyz(dacs, gogs, A)
% converts RGB DACS from a monitor to CIE XYZ
% dacs is a 3 by 1 matrix containing the RGB DACS (0-255)
% gogs is a 2 by 1 matrix containing the gamma and gain
% A is a 3 by 3 matrix to transform RGB to XYZ
```

11.21   *xyz2rgb.m* (Box 21)

```
% function [dacs] = xyz2rgb(XYZ, gogvals, A)
% converts XYZ to RGB DACS for a monitor
% XYZ is a 3 by 1 matrix containing the XYZ values
% gogvals is a 3 by 2 matrix containing the gamma and gain
% for each of the three channels
% A is a 3 by 3 matrix to transform RGB to XYZ
```

11.22   *compigog* (Box 22)

```
% function [dacs] = compgog(gogs,rgb)
% computes the normalized RGB values
% from the linearized RGB values
% for a given set of gog values
% gog is a 2 by 1 matrix that contains the gamma and gain
% dacs is an n by 1 matrix that contains the RGB values
% rgb is an n by 1 matrix of linearized RGB values
```

11.23   *getlincam.m* (Box 23)

```
% function [out] = getlincam(p,RGB,graphs)
% function to compute polynomial fits for camera
% grey-scale data. The inputs are p (a set of n by
% 1 mean reflectance values) and RGB ( a set of 3 by
% n RGB triplets). If graphs is set to 'on' then
% a plot fo the fits is generated
```

11.24   *lincam* (Box 24)

```
% function [RGBout] = lincam(caldata,RGB)
% computes linearized camera values using
% polynomial transforms obtained from getlincam
% caldata is a 3 by 4 matrix produced from getlincam.m
% RGB is an n by 3 matrix of RGB values (in range 0-255)
```

11.25   *gettrc* (Box 25)

```
% function [p] =; gettrc(dig,R,W,Solid,n,graphs)
% gettrc function to compute the trc for an ink
% function [p] =; gettrc(dig,R,W,Solid,n,graphs);
% dig is an 1 by n matrix of target area coverages
% R is an n by m matrix of measured reflectance values
% W is a 1 by m matrix of reflectance for the white substrate
% Solid is a 1 by m matrix of reflectance for the solid ink
% n is a free parameter > 0
% graphs =; 'on' for graphical display
% p is a matrix containing the coefficients of a polynomial
  to
% relate target coverage to actual coverage
```

11.26 *r2xyz* (Box 26)

```
% function [P] =; xyz2r(XYZ, obs)
% estimates refletcance P from XYZ trimulus values
% matrix XYZ must be n by 3
% illuminants A, C, D50, D55, D65, D75, F2, F7, F9
% set obs to 'd65_64 for D65 and 1964, for example
% matrix P is returned as an n by 31 matrix
```

References

Aleksander, I. and Morton, H. (1991). *An Introduction to Neural Computing*. Chapman & Hall.

Anton, H. (1994). *Elementary Linear Algebra*. John Wiley & Sons.

ASTM (2001). E308-01, Standard Practice for Computing the Colors of Objects by Using the CIE System.

Bala, R. (1999). Optimization of the spectral Neugebauer model for printer characterization, *Journal of Electronic Imaging*, **8** (2), 156–166.

Bala, R. (2003). Device characterization, in *Digital Color Imaging Handbook*, G. Sharma (Ed.), CRC Press.

Bartleson, C.J. and Breneman, E.J. (1967). Brightness perception in complex fields, *Journal of the Optical Society of America*, **57**, 953–957.

Berns, R.S. (1993). The mathematical development of CIE TC 1-29 proposed colour difference equation: CIELCH, *Proceedings of the Seventh Congress of International Colour Association*, B, C19.1–C19.4.

Berns, R.S. (2000). *Billmeyer and Saltzman's Principles of Color Technology*. John Wiley & Sons.

Berns, R.S. and Katoh, N. (2002). Methods for characterizing displays, in *Colour Engineering*, P. Green and L.W. MacDonald (Eds). John Wiley & Sons.

Berns, R.S., Motta, R.J. and Gorzynski, M.E. (1993a). CRT colorimetry. Part I: theory and practice, *Color Research and Application*, **18**, 299–314.

Berns, R.S., Motta, R.J. and Gorzynski, M.E. (1993b). CRT colorimetry. Part II: metrology, *Color Research and Application*, **18**, 315–325.

Borse, G.J. (1997). *Numerical Methods with MATLAB: A Resource for Scientists and Engineers*. International Thomson Publishing.

Bowmaker, J. (2002). The retina, in *Signals and Perception: The Fundamentals of Human Sensation*, D. Roberts (Ed.). Palgrave.

Brill, M.H. (1979). Further features of the illuminant invariant trichromatic photosensor, *Journal of Theoretical Biology*, **78**, 305–308.

Carlson, G.E. (1998). *Signal and Linear System Analysis*. John Wiley & Sons.

Cheung, T.L.V. and Westland, S. (2002). Color camera characterization using artificial neural networks, *Proceedings of the Tenth Color Imaging Conference*, Arizona (USA), pp. 117–120.

Cheung, T.L.V., Westland, S., Connah, D.R. and Ripamonti, C. (2004). A comparative study of the characterization of colour cameras by means of neural networks and polynomial transforms. *Journal of Coloration Technology* (in press).

Chou, W., Lin, H., Luo, M.R., Rigg, B., Westland, S. and Nobbs, J.H. (2001). The performance of the new CIE lightness difference formula, *Journal of Coloration Technology*, **117**, 19–29.

Computational Colour Science Using MATLAB. By Stephen Westland and Caterina Ripamonti.
© 2004 John Wiley & Sons, Ltd: ISBN 0 470 84562 7

CIE (1986a). Publication No. S2, *CIE Standard Colorimetric Observers*. Central Bureau of the CIE, Vienna.

CIE (1986b). Publication No. 15.2, *Colorimetry*, 2nd edn. Central Bureau of the CIE, Vienna.

CIE (1987). Publication No. 17.4, *CIE International Lighting Vocabulary*. Central Bureau of the CIE, Vienna.

CIE (2001). Publication No. 142, *Improvement to Industrial Colour-Difference Evaluation*. Central Bureau of the CIE, Vienna.

Clarke, F.J.J., McDonald, R. and Rigg, B. (1984). Modification to the JPC79 Colour-difference formula, *Journal of the Society of Dyers and Colourists*, **100**, 117–148.

Connah, D., Westland, S. and Thomson, M.G.A. (2001). Recovering spectral information using digital camera systems, *Journal of Coloration Technology*, **117**, 309–312.

Cui, G., Luo, M.R., Rigg, B. and Li, W. (2001). Colour-difference evaluation using CRT colours. Part I: data gathering and testing colour-difference formulae, *Color Research and Application*, **26**, 394–402.

D'Zmura, M. and Iverson, G. (1993a). Color constancy. I. Basic theory of two-stage linear recovery of spectral descriptions for lights and surfaces, *Journal of the Optical Society of America*, **A10**, 2148–2165.

D'Zmura, M. and Iverson, G. (1993b). Color constancy. II. Results for two-stage linear recovery of spectral descriptions for lights and surfaces, *Journal of the Optical Society of America*, **A10**, 2166–2180.

D'Zmura, M. and Lennie, P. (1986). Mechanisms of colour constancy, *Journal of the Optical Society of America*, **A3**, 1662–1672.

Emmel, P. and Hersch, R.D. (2000). A unified model for color prediction of halftoned prints, *Journal of Imaging Science and Technology*, **44**, 351–359.

Fairchild, M.D. (1998). *Color Appearance Models*. Addison Wesley Longman.

Fairchild, M.D. and Lennie, P. (1992). Chromatic adaptation to natural and incandescent illuminants, *Vision Research*, **32**, 2077–2085.

Fairchild, M.D. and Reniff, L. (1995). Time course of chromatic adaptation for color-appearance judgements, *Journal of the Optical Society of America*, **A12**, 824–833.

Finlayson, G.D. and Süsstrunk, S. (2000). Performance of a chromatic adaptation transform based on spectral sharpening, *Proceedings of the IS&T Eighth Colour Image Conference (Scottsdale)*, pp. 49–55.

Green, P. (2002a). Colorimetry and colour difference, in *Colour Engineering*, P. Green and L.W. MacDonald (Eds). John Wiley & Sons.

Green, P. (2002b). Overview of characterization methods, in *Colour Engineering*, P. Green and L.W. MacDonald (Eds). John Wiley & Sons.

Green, P. (2002c). Characterizing hard copy printers, in *Colour Engineering*, P. Green and L.W. MacDonald (Eds). John Wiley & Sons.

Green, P. and MacDonald, L.W. (2002). *Colour Engineering*. John Wiley & Sons.

Hardeberg, J.Y. (1999). Acquisition and reproduction of colour images: colorimetric and multispectral approaches. *PhD Thesis*, Ecole Nationale Supérieur des Télécommunications.

Hardeberg, J.Y. (2001). *Acquisition and Reproduction of Colour Images: Colorimetric and Multi-spectral Approaches*. Dissertation.

Haykin, S. (1994). *Neural Networks, A Comprehensive Foundation*. Macmillan.

Heptinstall, A. (1999). The perception of achromatic lightness differences. *MPhil Thesis*, Keele University.

Hunt, R.W.G. (1952). Light and dark adaptation and the reception of color, *Journal of the Optical Society of America*, **42**, 190–199.

Hunt, R.W.G. (1998). *Measuring Colour*. Fountain Press.

Hurlbert, A. (1991). Deciphering the colour code, *Nature*, **349**, 191–193.

Iino, K. and Berns, R.S. (1998). Building color management modules using linear optimization I. Desktop color system, *Journal of Imaging Science and Technology*, **42** (1).

Imai, F.H. and Berns, R.S. (1999). Spectral estimation using trichromatic digital cameras, *Proceedings of the International Symposium on Multispectral Imaging and Color Reproduction*, pp. 42–49.

Johnson, A.J. (1996). Methods for characterizing colour printers, *Displays*, **16**, 193–202.

Johnson, A.J. (2002). Methods for characterizing colour scanners and digital cameras, in *Colour Engineering*, P. Green and L.W. MacDonald (Eds). John Wiley & Sons.

Jolliffe, I.T. (1986). *Principal Components Analysis*. Springer-Verlag.

Judd, D.B., MacAdam, D.L. and Wyszecki, G.W. (1964). Spectral distribution of typical daylight as a function of correlated colour temperature, *Journal of the Optical Society of America*, **54**, 1031–1040.

Kaiser, P.K. and Boynton, R.M. (1996). *Human Colour Vision*. Optical Society of America.

Kang, H.R. (1994). Applications of colour mixing models to electronic printing, *Journal of Electronic Imaging*, **3**, 276–287.

Kohonen, T. (1988). *Self-organization and Associative Memory*. Springer-Verlag.

Lam, K.M. (1985). Metamerism and Colour Constancy. *PhD Thesis*, Bradford University (UK).

Land, E.H. (1986). Recent advances in retinex theory, *Vision Research*, **26**, 7.

Land, E.H. and McCann, J.J. (1971). Lightness and the retinex theory, *Journal of the Optical Society of America*, **61**, 1–11.

Li, C.J. and Luo, M.R. (2001). The estimation of spectral reflectances using the smoothness constraint condition, *Proceedings of the IS&T Ninth Colour Image Conference (Scottsdale)*, in press.

Li, C.J., Luo, M.R. and Hunt, R.W.G. (2000). Revision of the CIECAM97s model, *Color Research and Application*, **25**, 260–266.

Li, C.J., Luo, M.R., Rigg, B. and Hunt R.W.G. (2002). CMC 2000 Chromatic Adaptation Transform: CMCCAT2000, *Color Research and Application*, **27**, 49–58.

Luo, M.R. (2002a). Development of colour-difference formulae, *Review of Progress in Coloration*, **32**, 28–39.

Luo, M.R. (2002b). The CIE 1997 colour appearance model: CIECAM97s, in *Colour Engineering*, P. Green and L.W. MacDonald (Eds). John Wiley & Sons.

Luo, M.R. (2003). Private Communication.

Luo, M.R., Cui, G. and Rigg, B. (2001). The development of the CIE 2000 colour-difference formula: CIEDE2000, *Color Research and Application*, **26**, 340–350.

Luo, M.R. and Hunt, R.W.G. (1998a). The structure of the CIE 1997 colour appearance model (CIECAM97s), *Color Research and Application*, **23**, 138–146.

Luo, M.R. and Hunt, R.W.G. (1998b). A chromatic adaptation transform and a colour inconstancy index, *Color Research and Application*, **23**, 154–158.

Maloney, L.T. (1986). Evaluation of linear models of surface spectral reflectance with small numbers of parameters, *Journal of the Optical Society of America*, **3**, 1673–1683.

Maloney, L.T. and Wandell, B.A. (1986). Color constancy: a method for recovering surface spectral reflectance, *Journal of the Optical Society of America*, **A3** (1), 29–33.

Marchand, P. (1999). *Graphics and GUIs with MATLAB*. CRC Press.

McCann, J.J., McKee, S.P. and Taylor, T.H. (1976). Quantitative studies in retinex theory. A comparison between theoretical predictions and observer responses to the 'Colour Mondrian' experiments, *Vision Research*, **16**, 445–458.

McDonald, R. (1997a). *Colour Physics for Industry*. The Society of Dyers and Colourists.

McDonald, R. (1997b). Recipe prediction for textiles, in *Colour Physics for Industry*, McDonald, R. (Ed.). Society of Dyers and Colourists.

McDowell, D.Q. (2002). Standards activities for colour imaging, in *Colour Engineering*, P. Green and L.W. MacDonald (Eds). John Wiley & Sons.

Moroney, N. (2000). Usage guidelines for CIECAM97s, *Proceedings of PICS*, pp. 164–168.

Morovic, P. (2002). Colour gamut mapping, in *Colour Engineering*, P. Green and L.W. MacDonald (Eds). John Wiley & Sons.

Nayatani, Y., Takahama, K., Sobagaki, H. and Hashimoto, K. (1990). Color appearance model and chromatic adaptation transform, *Color Research and Application*, **15**, 210–221.

Nayatani, Y., Yano, T., Hashimoto, K. and Sobagaki, H. (1999). Proposal of an abridged colour-appearance model, *Color Research and Application*, **24**, 422–438.

Nobbs, J.H. (1985). Kubelka–Munk theory and the prediction of reflectance, *Review of Progress in Coloration*, **15**, 66–75.

Nobbs, J.H. (1997). Colour-match prediction for pigmented materials, in *Colour Physics for Industry*, McDonald, R. (Ed). Society of Dyers and Colourists.

Owens, H.C. (2002a). Private Communication.

Owens, H.C. (2002b). Spatiochromatic processing in humans and machines. *PhD Thesis*, Derby University (UK).

Poynton, C. (2002). Frequently asked questions about gamma, `http://www.inforamp.net/~poynton/notes/colour_and_gamma/GammaFAQ.html`.

Pratt, W.K. (1978). *Digital Image Processing*. John Wiley & Sons.

Press, W.H., Flannery, B.P., Teulolsky, S.A. and Vetterling, W.T. (1993). *Numerical Recipes in C: The Art of Scientific Computing*. Cambridge University Press.

Rich, D. (2002). Instruments and methods for colour measurement, in *Colour Engineering*, P. Green and L.W. MacDonald (Eds). John Wiley & Sons.

Roberts, D. (2002). *Signals and Perception: The Fundamentals of Human Sensation*. Palgrave.

Rumelhart, D.E. and McClelland, J.L. (1986). *Parallel distributed processing*, Vols I and II. MIT Press.

Sharma, G. (2002). Comparative evaluation of color characterization and gamut of LCDs versus CTs, *Proceedings of SPIE, Color Imaging: Device-Independent Color, Color Hardcopy, and Applications VII*, R. Eschbach and G. Marcu (Eds), pp. 177–186.

Smith, K.J. (1997). Colour-order systems, colour spaces, colour difference and colour scales, in *Colour Physics for Industry*, R. McDonald (Ed.). The Society of Dyers and Colourists.

Stearns, E.I. and Stearns, R.E. (1988). An example of a method for correcting radiance data, *Color Research and Application*, **13**, 257–259.

Stevens, J.C. and Stevens, S.S. (1963). Brightness function: effects of adaptation, *Journal of the Optical Society of America*, **53**, 375–385.

Sueeprasan, S. (2003). Evaluation of colour appearance models and indices for evaluation of daylight simulators to provide predictable cross-media colour reproduction. *PhD Thesis*, Derby University (UK).

Terstiege, H. (1972). Chromatic adaptation: a state-of-the-art report, *Color Research and Application*, **1**, 19–23.

Thomson, M.G.A. and Westland, S. (2002). Color-imager calibration by parametric fitting of sensor responses, *Color Research and Application*, **26** (6), 442–449.

Tominaga, S. and Wandell, B.A. (1990). Component estimation of surface spectral reflectance, *Journal of the Optical Society of America*, **A7**, 312–317.

Venable, W.H. (1989). Accurate tristimulus values from reflectance data, *Color Research and Application*, **14**, 260–267.

Viggiano, J.A.S. (1990). Modeling the color of multi-colored halftones, *Proceedings of TAGA*, pp. 44–62.

Wandell, B.A. (1995). *Foundations of Vision*. Sinauer Associates Incorporated.

Westland, S. (2002). Functional colour vision, in *Signals and Perception: The Fundamentals of Human Sensation*, D. Roberts (Ed.). Palgrave.

Westland, S. Shaw, A.J. and Owens, H.C. (2000). Colour statistics of natural and man-made surfaces, *Sensor Review*, **20** (1), 50–55.

Williams, D., MacLeod, D.I.A. and Hayhoe, M. (1981). Punctate sensitivity of the blue sensitive mechanism, *Vision Research*, **21**, 1357–1375.

Wyszecki, G. and Stiles, W.S. (1982). *Color Science: Concepts and Methods, Quantitative data and formulae*. John Wiley & Sons.

Xu, H., Luo, M.R. and Rigg, B. (2003). The variation of the quality of D65 and D50 daylight simulators, *Journal of Coloration Technology* (in press).

Yule, J.A.C. (1967). *Principles of Color Reproduction – Applied to Photomechanical Reproduction, Colour Photography, and the Ink, Paper, and Other Related Industries*. John Wiley & Sons.

Yule, J.A.C. and Nielsen, W.J. (1951). The penetration of light into paper and its effect on halftone reproduction, *Proceedings of TAGA*, pp. 65–76.

Index

adapted white 86
additive primaries 5, 8
adopted white 86
ANLAB colour space 50
ANSI IT8.7 charts 128
artificial neural networks 111, 130, 136, 143, 156

basis functions 166, 176
brightness 82

Cartesian coordinates 50, 65
channel balancing 136
chroma 82
chromatic adaptation 81, 83
  degree of 86, 99, 103
  transform 81
chromatic sharpening 85
CIE System
  chromaticity coordinates 8, 35
  chromaticity diagram 35, 46
  CIE 1976 UCS 51, 63
  CIELAB 9, 50–52, 79, 92
  CIELUV 9, 50–52
  Colour-matching functions 6, 28
  illuminants 29, 44, 45
  illuminant white points 45
  limitations of 8–9
  standard observer 8, 28
  tristimulus values 5–7, 27
  tristimulus values (computing) 41
CIE94 equation 49, 56, 73, 79
CIECAM97s 82, 93–96
CIECAT94 82, 86–88
CIECAM97s 82, 93–96
CMCCAT97 82, 89–90, 96, 104
CMCCAM2000 82, 96

CMCCAT2000 82, 90–91, 96, 100
CIEDE2000 equation 10, 49, 57, 75, 79
CMC equation 10, 49, 55, 71, 79
colorimeter (split-field) 5
colour
  appearance 7–8, 54, 81–82, 85, 104, 164
  appearance model 81, 92
  constancy 10, 81, 164
  communication 11
  difference 9–10, 49, 52
  difference test data 78
  memory 82
Colour Measurement Committee 10, 55
colourfulness 82
continuous-tone printer 155
corresponding colours 82
CRT spatial independence 115
CRT channel independence 115

daylight simulators 29
device calibration 111
device characterization 111, 115
device linearization 113, 128, 145
device-independent representation 114, 129, 146
dot gain 142, 145

eigenvectors 167

Fourier analysis 185

gamma 112–114
gamut 8, 36
GOG model 111–114
Grassman's law 8
grey world assumption 170

half-tone printers 142, 150

hue
  computing differences 53, 75
  difference descriptors 54
Hunt effect 82, 93
HunterLab 50

JPC79 equation 55

von Kries 84, 90, 164
Kubelka-Munk model 111, 142, 147

lagrange polynomial 30, 33
LCD characterization 116
Levenberg-Marquardt optimization 157
lightness 82
linear algebra
  diagonal transform 84, 90
  identity matrix 14–15
  inverse of a matrix 16
  linear transform 16
  matrix, definition of 13
  matrix augmentation 20
  matrix multiplication 15
  polynomial transform 18, 130, 137
  simultaneous equations 14–15
  transpose of a matrix 16
  van der monde matrix 30, 40

Macbeth ColorChecker 127, 129, 136
Macbeth ColorChecker DC 127, 136
masking model 159
matrix mixing 128
MATLAB
  abs command 185
  advantages 2–4
  angle command 185
  atan2 command 67
  backslash command 22, 30
  clear command 22
  cond command 25
  double command 135
  fft command 185
  fminsearch command 121
  functions 25
  imread command 135
  init command 157
  interp1 command 32–33
  introduction 19–26
  inv command 19, 23
  length command 38
  load command 22
  M-files 3, 25
  newff command 156
  ones command 41
  pinv command 23

polyfit command 30, 133
polyval command 30, 152
reshape command 135
save command 22
sim command 158
size command 22
svd 167
svds 167, 172
train command 157
van der Monde matrix 30
monitor characterization example 117
multispectral Imaging 163, 171
Murray-Davies model 142, 145

Neugebauer model 142, 146, 152
Neville's algorithm 33
Nyquist limit 186

opto-electronic transfer function 112, 116
orthogonality, condition of 177

polar coordinates 50, 55, 65
principal components analysis 167, 172, 176

reflectance
  extrapolation 33
  estimation from camera responses 171, 183
  estimation from cone responses 164
  estimation from XYZ values 179
  interpolation 31, 39
  Fourier analysis of 166
  linear models of 165, 169
  measurement 28
  spectra for natural surfaces 32, 169
  spectrophotometer 4, 28
retinex theory 164

scattering 148
Simplex algorithm 115
simultaneous contrast 92
singular value decomposition 167
spectral bandpass correction 35, 37
spectral frequency 166, 185
spectral imaging 163
Stearns-Stearns bandpass correction 35

tone reproduction curves 152
toolbox functions
  car2pol 65, 75
  cband 37
  cie00de 75
  cie94de 73
  ciecam97s 104
  cielabde 68

cmccat00 100
cmccat97 98
cmcde 71
compgog 117, 121
compigogs 124
dhpolarity 70
getlincam 131
gettrc 150
gogtest 120
lab2xyz 61
lincam 134
pinterp 33, 39, 41, 46
pol2car 67, 75
plocus 45–46
r2xyz 41, 44
rgb2xyz 122

testgog 117
xyz2rgb 123
xyz2lab 58
xyz2luv 63

vision
    cone responses 4, 164
    photopic/scotopic response 4
    spectral sensitivity 5
    cortex 4
    cone spatial distribution 5

weights, tables of 34
weights (ASTM) 34, 39, 41, 44
weights.mat file 45
white-point balancing 127